W9-CSI-499

Quantitative Applications in the Social Sciences

A SAGE PUBLICATIONS SERIES

Quantitative Applications in the Social Sciences

A SAGE PUBLICATIONS SERIES

Series/Number 07–153

AGENT-BASED MODELS

Nigel Gilbert
University of Surrey, Guildford, UK

SAGE Publications
Los Angeles ▪ London ▪ New Delhi ▪ Singapore

For information:

Sage Publications, Inc.
2455 Teller Road
Thousand Oaks, California 91320
E-mail: order@sagepub.com

Sage Publications India Pvt. Ltd.
B 1/I 1 Mohan Cooperative Industrial Area
Mathura Road, New Delhi 110 044
India

Sage Publications Ltd.
1 Oliver's Yard
55 City Road
London EC1Y 1SP
United Kingdom

Sage Publications Asia-Pacific Pte. Ltd.
33 Pekin Street #02-01
Far East Square
Singapore 048763

Printed in the United States of America

Library of Congress Cataloging-in-Publication Data

Gilbert, G. Nigel.
   Agent-based models / Nigel Gilbert.
      p. cm.—(Quantitative applications in the social sciences; 153)
   Includes bibliographical references and index.
   ISBN 978-1-4129-4964-4 (pbk.)
      1. Social sciences—Statistical methods. 2. Game theory. 3. Computer simulation. I. Title.
   HA29.G527 2008
   519.3—dc22

                                                    2007014599

This book is printed on acid-free paper.

07   08   09   10   11   10   9   8   7   6   5   4   3   2   1

Acquisitions Editor:	Vicki Knight
Associate Editor:	Sean Connelly
Production Editor:	Melanie Birdsall
Copy Editor:	Liann Lech
Typesetter:	C&M Digitals (P) Ltd.
Proofreader:	Scott Oney
Indexer:	Sylvia Coates
Cover Designer:	Candice Harman
Marketing Manager:	Stephanie Adams

CONTENTS

SERIES EDITOR'S INTRODUCTION

There are two general approaches to the study of social behavior: Collect observational, survey, or other forms of data and analyze them, possibly by estimating a model; or begin from a theoretical understanding of certain social behavior, build a model of it, and then simulate its dynamics to gain a better understanding of the complexity of a seemingly simple social system. Most of the books in Sage's Quantitative Applications in the Social Sciences series take the first approach; *Agent-Based Models*, as one of the very few exceptions, belongs in the second tradition.

Many of the statistical methods applied in the social sciences have had a long history of development. For example, regression methods have their foundation in Galton's groundbreaking work in the 19th century. The origin of agent-based models is often attributed to Von Neumann because he developed his namesake machines in the late 1940s even though Von Neumann machines are not agent-based models or even models of any kind at all; they are theoretical constructs that paved the foundation for the construction and modeling of artificial life. Agent-based models have become increasingly popular in the past decade or so in economics, political science, sociology, and some of the other social sciences.

What is so unique about this analytical approach? Take, for example, Thomas Schelling's classical work published in the *Journal of Mathematical Sociology* (1971). He built a very simple dynamic model of segregation by moving pennies and dimes on a chessboard according to certain simple rules. The pennies and dimes represented agents of two different social groups who resided in squares (houses or lots) in the chessboard (city). The surprising finding was that, although each agent was tolerant of neighbors from a different group (or mildly preferred own group), the population (in the entire chessboard-city) ended up in severe segregation. Schelling's simple agent-based model reveals some profound relations between micro-level individual motives and macro-level social behavior.

Nigel Gilbert, an international authority on agent-based models and editor of the *Journal of Artificial Societies and Social Simulation*, is well suited for writing a book on the topic for the series. In this book, Gilbert explains the fundamental principles of and discusses the software for agent-based models by using examples throughout. He writes with clarity and precision, important for making an unfamiliar topic accessible to researchers who are on the familiar territory of the usual regression type of methods. This book is timely in filling a significant gap in the series that I as series editor had hoped to have filled, and it should facilitate the diffusion of a fundamentally useful method that we as social scientists all should know.

—Tim F. Liao
Series Editor

PREFACE

Agent-based modeling is a form of computational simulation. Although simulation as a research technique has had a very important part to play in the natural sciences for decades in disciplines from astronomy to biochemistry, until recently it was neglected in the social sciences. This may have been because a computational approach that respected the particular needs of the social sciences was lacking. However, in the early 1990s, the value of agent-based modeling began to be realized, and since then, the number of studies that have used agent-based modeling has grown very rapidly, and examples are beginning to appear in core journals (e.g., Gilbert & Abbott, 2005).

Agent-based modeling is particularly suited to topics where understanding processes and their consequences is important. In essence, one creates a computer program in which the actors are represented by segments of program code, and then runs the program, observing what it does over the course of simulated time. There is a direct correspondence between the actors being modeled and the agents in the program, which makes the method intuitively appealing, especially to those brought up in a generation used to computer games. Nevertheless, agent-based modeling stands beside mathematical and statistical modeling in terms of its rigor. Like equation-based modeling, but unlike prose, agent-based models have to be complete, consistent, and unambiguous if they are to be capable of being executed on a computer. On the other hand, unlike most mathematical models, agent-based models can include agents that are heterogeneous in their features and abilities, can model situations that are far from equilibrium, and can deal directly with the consequences of interaction between agents.

Because it is a new approach, there are few courses yet available to teach the skills of agent-based modeling, although the number is increasing rapidly, and few texts directed at the interested social scientist (Degushi, 2004; Gilbert & Terna, 2000; Gilbert & Troitzsch, 2005; Grimm & Railsback, 2005; Liebrand, Nowak, & Hegselmann, 1998; Macy & Willer, 2002; Schweitzer, 2003; Tesfatsion & Judd, 2006). This short book provides an introduction to the subject, emphasizing the decisions that a modeler needs to make when selecting agent-based modeling as an appropriate method, and offering some tips on how to proceed. It is aimed at practicing social scientists and graduate students. It has been used as the recommended reading on agent-based modeling for a graduate-level module or doctoral program in computational social science, and it is also suitable as background reading in postgraduate courses on advanced social research methods.

A knowledge of and experience with computer programming in any language would be helpful but is not essential to understand the book.

The book concludes with a list of printed and Web resources, glossary, and reference section. Because the field is growing so rapidly, it has been possible to mention only a few examples of current research and some textbooks that provide more detail on particular topics. There is much more that could have been cited if there had been space. In particular, the book only mentions briefly two closely linked areas: network models and game theory models, both of which are covered in much more detail in other Sage volumes such as Scott's *Social Network Analysis* (2000) and Fink, Gates, and Humes's *Game Theory Topics* (1998).

ACKNOWLEDGMENTS

This book is borne of some 15 years of building agent-based models, both large and small, and in domains ranging from science policy to anthropology. What I know about agent-based modeling has benefited immeasurably from the advice and companionship of many, including, in alphabetical order, Andrew Abbott, Petra Ahrweiler, Robert Axelrod, Rob Axtell, Riccardo Boero, François Bousquet, Cristiano Castelfranchi, Edmund Chattoe, Claudio Cioffi-Revilla, Rosaria Conte, Guillaume Deffuant, Bruce Edmonds, Gusz Eiben, Wander Jager, David Lane, Scott Moss, Gilbert Peffer, Andreas Pyka, Juliette Rouchier, Stephan Schuster, Luc Steels, Klaus Troitzsch, Paul Vogt, and Lu Yang. I thank Riccardo Boero, Lars-Eric Cederman, Lynne Hamill, Ken Kahn, Luis R. Izquierdo, Michael Macy, Tim Liao, Lu Yang, and two anonymous reviewers for their detailed and constructive comments on drafts of the manuscript.

ACKNOWLEDGMENTS

AGENT-BASED MODELS

Nigel Gilbert
University of Surrey, Guildford, UK

1. THE IDEA OF AGENT-BASED MODELING

Agent-based modeling is a new analytical method for the social sciences, but one that is quickly becoming popular. This short book explains what agent-based modeling is. It also warns of some dangers and describes typical ways of doing agent-based modeling. Finally, it offers a range of examples from many of the social sciences.

This first chapter begins with a brief overview of agent-based modeling before contrasting it with other, perhaps more familiar forms of modeling and describing several examples of current agent-based modeling research. Chapter 2 goes into more detail, considering a range of methodological and theoretical issues and explaining what "agents" are. Chapter 3 dives into the specifics of building agent-based models, reviewing available software platforms and showing step by step how one can build an agent-based model using one of these. Chapter 4 provides some practical advice about designing agent-based models, using them in social science research, and publishing articles based on agent-based modeling. Finally, Chapter 5 discusses the future of agent-based modeling research and where advances are likely to be made. The book concludes with a list of resources useful to agent-based modelers on the Web and in print.

Agent-based simulation has become increasingly popular as a modeling approach in the social sciences because it enables one to build models where individual entities and their interactions are directly represented. In comparison with variable-based approaches using structural equations, or system-based approaches using differential equations, agent-based simulation offers the possibility of modeling individual heterogeneity, representing explicitly agents' decision rules, and situating agents in a geographical or another type of space. It allows modelers to represent in a natural way multiple scales of analysis, the emergence of structures at the macro or societal level from individual action, and various kinds of adaptation and learning, none of which is easy to do with other modeling approaches.

1.1 Agent-Based Modeling

Formally, agent-based modeling is a *computational* method that enables a researcher to create, analyze, and *experiment* with *models* composed of *agents* that interact within an *environment*. Let us consider each of the italicized terms in this definition.

1.1.1 A Computational Method

First, agent-based modeling is a form of *computational social science*. That is, it involves building models that are computer programs. The idea of modeling is familiar in most of the social sciences: One creates some kind of simplified representation of "social reality" that serves to express as clearly as possible the way in which one believes that reality operates. For example, if one has a dependent variable and one or more independent variables, a regression equation serves as a model of the relationship between the variables. A network of nodes and edges can model a set of friendships. Even an ordinary language description of a relationship, such as that between the strength of protection of intellectual property rights and the degree of innovation in a country, can be considered a model, albeit a simple and rather unformalized one.

Computational models are formulated as computer programs in which there are some inputs (somewhat like independent variables) and some outputs (like dependent variables). The program itself represents the processes that are thought to exist in the social world (Macy & Willer, 2002). For example, we might have a theory about how friends influence the purchasing choices that consumers make. As we shall see, we can create a program in which there are individuals ("agents") that buy according to their preferences. The outcome is interesting because what one agent buys will influence the purchasing of a friend, and what the friend buys will influence the first agent. This kind of mutual reinforcement is relatively easy to model using agent-based modeling.

One of the advantages of computational modeling is that it forces one to be precise: Unlike theories and models expressed in natural language, a computer program has to be completely and exactly specified if it is to run. Another advantage is that it is often relatively easy to model theories about processes, for programs are all about making things within the computer change. If the idea of constructing computational models reminds you of computer games, especially the kind where the player has a virtual world to build, such as The Sims (http://thesims.ea.com/), that is no accident. Such games can be very close to computational modeling, although in order to make them fun, they often have fancier graphics and less social theory in them than do agent-based models.

1.1.2 Experiments

Whereas in physics and chemistry and some parts of biology, experimentation is the standard method of doing science, in most of the social sciences, conducting experiments is impossible or undesirable. An experiment consists of applying some *treatment* to an isolated system and observing what happens. The treated system is compared with another otherwise equivalent system that receives no treatment (the control). The great advantage of experiments is that they allow one to be sure that it is the treatment that is causing the observed effects, because it is only the treatment that differs between the treated and the control systems and the systems are isolated from other potential causes of change. However, with social systems, isolation is generally impossible, and treating one system while not treating the control is often ethically undesirable. Therefore, it is not surprising that experiments, despite the potential clarity of their results, are rarely used by social scientists.

A major advantage of agent-based modeling is that the difficulties in ensuring isolation of the human system and the ethical problems of experimentation are not present when one does experiments on virtual or computational systems. An experiment can be set up and repeated many times, using a range of parameters or allowing some factors to vary randomly. Of course, carrying out experiments with a computational model of some social phenomenon will yield interesting results only if the model behaves in the same way as the human system or, in other words, if the model is a good one, and one may not know whether that is the case. So, experimentation on models is not a panacea.

The idea of experimenting on models rather on the real system is not novel. For example, when architects put a model tower block in a wind tunnel to investigate its behavior in high winds, they are experimenting on the model for just the same reasons as social scientists might want to experiment on their models: The cost of experimenting on a real tower block is too high. Another reason for experimenting with models is that this may be the only way to obtain results. Deriving the behavior of a model analytically is usually best because it provides information about how the model will behave given a range of inputs, but often an analytical solution is not possible. In these cases, it is necessary to experiment with different inputs to see how the model behaves. The model is used to *simulate* the real world as it might be in a variety of circumstances.

1.1.3 Models

Computational social science is based on the idea of constructing models and then using them to understand the social world (Sawyer, 2004). Models have a long history in the social sciences—much longer than the use of computers—but came to the fore when statistical methods began to be used to analyze large amounts of quantitative data in economics and demography. A model is intended to represent or simulate some real, existing phenomenon, and this is called the *target* of the model. The two main advantages of a model are that it succinctly expresses the relationships between features of the target, and it allows one to discover things about the target by investigating the model (Carley, 1999).

One of the earliest well-known social science models is the Phillips (1950) hydraulic model of the economy in which water flowing through interconnected glass pipes and vessels is used to represent the circulation of money. This model can still be admired at the Science Museum, London (http://en.wikipedia.org/wiki/MONIAC_Computer). The effect of changing parameters such as the interest rate can be investigated by changing the rate of flow of water through the pipes.

Models come in several flavors, and it is worth listing some of these to clarify the differences:

• *Scale models* are smaller versions of the target. Together with the reduction in size is a systematic reduction in the level of detail or complexity of the model. So, for example, a scale model of an airplane will be the same shape as its target, but probably would not show the electronic control systems or possibly even the engines of the real plane. Similarly, a scale model of a city will be much smaller than the real city and may model only two dimensions (the distances between buildings, but not the heights of buildings, for instance). When drawing conclusions about the target by studying the model, one needs to bear in mind that the results from the model will need to be scaled back up to the target's dimensions, and that it is possible that some of the features not modeled may affect the validity of the conclusions.

• An *ideal-type* model is one in which some characteristics of the target are exaggerated in order to simplify the model. For example, an idealized model of a stock market may assume that information flows from one trader to another instantaneously, and an idealized model of traffic may assume that drivers never get lost. The idealization has the effect of removing one or more complicating factors from the model, and if these have negligible effects on how the model works, the model will remain useful for drawing conclusions about the target.

• *Analogical models* are based on drawing an analogy between some better understood phenomenon and the target. The most famous example is the billiard ball model of atoms, but there are also social science examples such as the computer model of the mind (Boden, 1988) and the garbage can model of organizations (Cohen, March, & Olsen, 1972). Such models are useful because well-established results from the analogy can be carried over and applied to the target, but of course the validity of these depends on the adequacy of the analogy.

These are not mutually exclusive categories; it is possible, and indeed common, for a model to be a scale model and an analogy (for example, the hydraulic model of the economy mentioned above is such a combination).

Some models fall into a fourth category that is somewhat different, but also commonly encountered in the social sciences; these are often called *mathematical* or *equation-based* models. Examples are the structural equation models of quantitative sociology and the macroeconomic models of neoclassical economics. These models specify relationships between variables, but unlike models in the other three categories, they do not imply any kind of analogy or resemblance between the model and the target. Usually, the success of a mathematical model is indicated by the degree to which some data fit the equation, but the form of the equation itself is of little interest or consequence. For example, the Cobb-Douglas "production function" is a mathematical model of how manufactured outputs are related to inputs (Cobb & Douglas, 1928):

$$Y = AL^{\alpha}K^{\beta},$$

where $Y =$ output, $L =$ labor input, $K =$ capital input, and A, α, and β are constants determined by technology. The form of this equation was derived from statistical evidence, not by theorizing about the behavior of firms. Although mathematical models have been very successful in some parts of the social sciences in clarifying the relationships between variables, they are often not very useful in helping to understand *why* one variable is related to another, or in other words, in expressing ideas about process and mechanism, where the other types of models are generally more helpful.

1.1.4 Agents

Agent-based models consist of agents that interact within an environment. Agents are either separate computer programs or, more commonly, distinct parts of a program that are used to represent social actors—individual people, organizations such as firms, or bodies such as nation-states. They are programmed to react to the computational environment in which they are located, where this environment is a model of the real environment in which the social actors operate.

As will be seen later, a crucial feature of agent-based models is that the agents can interact, that is, they can pass informational *messages* to each other and act on the basis of what they learn from these messages. The messages may represent spoken dialogue between people or more indirect means of information flow, such as the observation of another agent or the detection of the effects of another agent's actions. The possibility of modeling such agent-to-agent interactions is the main way in which agent-based modeling differs from other types of computational models.

1.1.5 The Environment

The environment is the virtual world in which the agents act. It may be an entirely neutral medium with little or no effect on the agents, or in other models, the environment may be as carefully crafted as the agents themselves. Commonly, environments represent geographical spaces, for example, in models concerning residential segregation, where the environment simulates some of the physical features of a city, and in models of international relations, where the environment maps states and nations (Cederman, 1997). Models in which the environment represents a geographical space are called *spatially explicit*. In other models, the environment could be a space, but one that represents not geography but some other feature. For example, scientists can be modeled in "knowledge space" (Gilbert, Pyka, & Ahrweiler, 2001). In these spatial models, the agents have coordinates to indicate their location. Another option is to have no spatial representation at all but to link agents together into a network in which the only indication of an agent's relationship to other agents is the list of the agents to which it is connected by network links (Scott, 2000).

To make these definitions somewhat more concrete, in the next section we shall introduce some examples of agent-based models in terms of these concepts.

1.2 Some Examples

Agent-based models are of value in most branches of social science. The models that are briefly described in the rest of this section have been chosen to illustrate the diversity of the problem areas where they have been used productively.

1.2.1 Urban Models

In 1971, Thomas Schelling (1971, 1978; see also Sakoda, 1971) proposed a model to explain observed racial segregation in American cities. The model is a very abstract one as originally conceived, but it has been influential in recent work on understanding the persistence of segregation not only in the United States but also in other urban centers. The model is based on a regular square grid of cells representing an urban area on which agents, representing households, are placed at random. The agents are of two kinds (let us call them "reds" and "greens"). Each cell can hold only one household agent at a time, and many cells are empty. At each time step, each household surveys its immediate neighbors (the eight cells surrounding it) and counts the fraction of households that are of the other color. If the fraction is greater than some constant threshold "tolerance" value, that is, there are more than a fixed proportion of reds surrounding a green, or greens surrounding a red, that household considers itself to be "unhappy" and decides to relocate. It does so by moving to some vacant cell on the grid.

At the next time step, the newly positioned household may tip the balance of "tolerance" of its neighbors, causing some of them to become "unhappy," and this can result in a cascade of relocations. For levels of the tolerance threshold at or above about 0.3, an initially random distribution of households segregates into patches of red and green, with households of each color clustering together (Figure 1.1). The clustering occurs even when households "tolerate" living adjacent to a majority of neighbors of the other color, which Schelling interpreted as indicating that even quite low degrees of racial prejudice could yield the strongly segregated patterns typical of U.S. cities in the 1970s.

The Schelling model has been influential for several reasons (Allen, 1997). First, the outcome—clusters of households of the same color—is surprising and not easily predictable just from considering the individual agents' movement rule. Second, the model is very simple and has only one parameter, the "tolerance" threshold. It is therefore easy to understand. Third, the emergent clustering behavior is rather robust. The same outcomes are obtained for a wide range of tolerance values, for a variety of movement rules (e.g., the household agent could select a new cell at

Figure 1.1   The Schelling Model at the Start (Left) and After Equilibrium Has Been Reached (Right), With a Uniform Tolerance of 0.3

SOURCE: Wilensky, U. (1998). NetLogo Segregation model. http://ccl.northwestern.edu/netlogo/models/Segregation. Center for Connected Learning and Computer-Based Modeling, Northwestern University, Evanston, IL.

random, or use a utility function to select the most preferred cell, or take into account affordability if cells are priced, and so on) and for different rules about which neighbors to consider (e.g., those in the eight surrounding cells; the four cells to the north, east, south, and west; or a larger ring two or more cells away) (Gilbert, 2002). Fourth, the model immediately suggests how it could be tested with empirical data (Benenson, Omer, & Hatna, 2002; Bruch & Mare, 2006; Clark, 1991; Pancs & Vriend, 2004; Zhang, 2004), although in practice it has proved quite difficult to obtain reliable and extensive data on household location preferences to calculate ratings of "unhappiness." The advantages of the Schelling model over others that had been previously proposed, which were based on equations relating migration flows and the relative values of residential properties (e.g., Fotheringham & O'Kelly, 1989), are that the number of parameters to be estimated is lower and that the model is simple to simulate and analyze. Current work has focused on making the model more concrete, replacing the abstract square grid with actual urban geographies and adding further factors, such as the affordability of the locations to which households want to move.

1.2.2   Opinion Dynamics

Another interesting group of models with potentially important implications is concerned with understanding the development of political opinions, for example, with explaining the spread of extremist opinions within a population. We shall review just one such study, although there are a number that explore the consequences of different assumptions and opinion transmission mechanisms (Deffuant, 2006; Deffuant, Amblard, & Weisbuch, 2002; Hegselmann & Krause, 2002; Lorenz, 2006; McKeown & Sheehy, 2006; Stauffer, Sousa, & Schulze, 2004). Deffuant et al. (2002) ask,

> How can opinions, which are initially considered as extreme and marginal, manage to become the norm in large parts of a population? Several examples in world history show that large communities can more or less suddenly switch globally to one extreme opinion, because of the influence of an initially small minority. Germany in the thirties is a particularly dramatic example of such a process. In the last decades, an initial minority of radical Islamists managed to convince large populations in Middle East countries. But one can think of less dramatic processes, like fashion for instance, where the behavior of minority groups, once considered as extremist, becomes the norm in a large part of the population (it is the case of some gay way of dressing for instance). On the other hand, one can also find many examples where a very strong bipolarization of the population takes place, for instance the Cold War period in Europe. In these cases, the whole population becomes extremist (choosing one side or the other).

In Deffuant et al.'s model, agents have an opinion (a real number between −1.0 and +1.0) and a degree of doubt about their opinion, called *uncertainty* (a positive real number). An agent's *opinion segment* is defined as the band centered on the agent's opinion, spreading to the right and left by the agent's value for uncertainty. Agents interact randomly. When they meet, one agent may influence the other if their opinion segments overlap. If the opinion segments do not overlap, the agents are assumed to be so different in their opinions that they have no chance of influencing each other. If an agent does influence another, the opinion of one agent (agent j) is affected by the opinion of another agent (agent i) by an amount proportional to the difference between their opinions, multiplied by the amount of overlap divided by agent i's uncertainty, minus one. The effect of this formula is that very uncertain agents influence other agents less than those that are certain (for full details, see Deffuant et al., 2002, Equations 1 to 6).

Every agent starts with an opinion taken from a uniform random distribution and with a common level of uncertainty, with the exception of a few extremists, those who have the most positive or negative opinions. The latter are given a low level of uncertainty, that is, the extremists are assumed to be rather certain about their extreme opinions. Under these conditions, extremism spreads, and eventually the simulation reaches a steady state with all agents joining the extremists at one or the other end of the opinion continuum. Restarting the simulation without the politically certain extremists, the population converges instead so that all agents share a middle view. Thus, the model shows that a few extremists with opinions that are not open to influence from other agents can have a dramatic effect on the opinions of the majority. This work has some implications for the development of terrorist movements, where a few extremists have been able to recruit support from substantial proportions of the wider population.

1.2.3 Consumer Behavior

Businesses are naturally keen to understand what influences their customers to buy their products. Although the intrinsic qualities of the product are usually important, so are the influence of friends and families, advertising, fashion, and a range of other "social" factors. To examine the often complex interactions between these, some researchers have started to use agent-based models in which the agents represent consumers. Among the first to report such a model were Janssen and Jager (1999), who explored the processes leading to "lock-in" in consumer markets. Lock-in occurs when one among several competing products achieves dominance so that it is difficult for individual consumers to switch to rival products. Commonly cited examples are VHS videotapes (dominating Betamax), the QWERTY keyboard (dominating other arrangements of the keys), Microsoft operating systems (dominating Apple and Linux), and so on. Janssen and Jager focus on the behavioral processes that lead to lock-in, and therefore they give their agents, which they call "consumats," decision rules that are psychologically plausible and carefully justified in terms of behavioral theories of, for example, social comparison and imitation.

Another example of modeling consumers is a study by Izquierdo and Izquierdo (2006) in which the authors consider markets such as the second-hand car market, where there is quality variability (different quality for different items) and quality uncertainty (it is difficult to know the quality of an item before buying it and using it). The study explores how quality variability can damage a market and affect consumer confidence. There are two agent roles: buyer and seller. Sellers sell products by calculating the minimum price they will accept, and buyers buy products by offering a price based on the expected quality of the product. The expected quality is based on experience, either just of the agent or accumulated from the agent's peers over its social network. There are a finite number of products in the system, buyers and sellers perform one deal per round, and the market is cleared every round as these deals are done. The social network is created by connecting pairs of agents at random, with a parameter used to adjust the number of connections, from completely connected to completely unconnected.

The authors found that without a social network, consumer confidence fell to the point where the market was no longer viable, whereas with a social network, the aggregation of the agent's own experience and the more positive collective experience of others (which is not so volatile) helped to maintain the market's stability. This shows how social information can aggregate group experience to a more accurate level and so reduce the importance of a single individual's bad experiences.

1.2.4   Industrial Networks

Most economic theory ignores the significance of links between firms, but there are many examples of industrial sectors where networks are of obvious importance. A well-known instance is the "industrial districts" of northern Italy, such as the textile production district, Prato. Industrial districts are characterized by large numbers of small firms clustered together in a small region, all manufacturing the same type of product, with strong, but varying, links between them. The links may be those of a supplier/ customer, a collaboration to share techniques, a financial link, or just a friendship or familial relationship (Albino, Carbonara, & Giannoccaro, 2003; Boero, Castellani, & Squazzoni, 2004; Borrelli, Ponsiglione, Iandoli, & Zollo, 2005; Brenner, 2001; Fioretti, 2001; Squazzoni & Boero, 2002).

Another example is the "innovation networks" that are pervasive in knowledge-intensive sectors such as biotechnology and information technology. The firms in these sectors are not always geographically clustered (although there tend to be concentrations in certain locations), but they do have strong networking relationships with other, similar firms, sharing knowledge, skills, and technology with them.

For example, Gilbert et al. (2001) developed a model of innovation networks in which agents have "kenes" that symbolize their stock of knowledge and expertise. The kenes are used to develop new products that are marketed to other firms in the model. However, a product can be produced only if its components are available to be purchased from other firms, and if some firm wants to buy it. Thus, at one level, the model is one of an industrial market with firms trading between each other. In addition, firms can improve their kenes either through internal research and development or through incorporating knowledge obtained from other firms by collaborative arrangements. The improved knowledge can be used to produce products that may sell better, or require fewer or more cheaply available components. At this level, the model resembles a population that can learn through a type of natural selection (see Section 5.2.2) in which firms that cannot find a customer cease trading, whereas the fittest firms survive, collaborate with other firms, and produce spin-offs that incorporate the best aspects of their "parents." For another example, see Pajares, Hernández-Iglesias, and López-Paredes (2004).

1.2.5 Supply Chain Management

Manufacturers normally buy components from other organizations and sell their products to distributors, who then sell to retailers. Eventually, the product reaches the user, who may not realize the complex interorganizational relationships that have had to be coordinated to deliver the product. Maximizing the efficiency of the supply chain linking businesses is increasingly important and increasingly difficult as products involve more parts, drawn more widely from around the world, and as managers attempt to reduce inventory and increase the availability of goods. Modeling supply chains is a good way of studying order fulfillment processes and investigating the effectiveness of management policies, and multiagent models are increasingly being used for this purpose.

A multiagent model fits well with the task of simulating supply chains because the businesses involved can be modeled as agents, each with its own inventory rules. It is also easy to model the flow of products down the chain and the flow of information, such as order volumes and lead times, from one organization to another. This was the approach taken by Strader,

Lin, and Shaw (1998), who described a model they built to study the impact of information sharing in divergent assembly supply chains. Divergent assembly supply chains are typical of the electronics and computer industries and are those in which a small number of suppliers provide materials and subcomponents (e.g., electronic devices) that are used to assemble a range of generic products (e.g., hard disk drives) that are then used to build customized products at distribution sites (e.g., personal computers). Strader et al. compared three order fulfillment policies: make-to-order, when production is triggered by customer orders; assembly-to-order, when components are built and held in stock, and only the final assembly is triggered by an order; and make-to-stock, when production is driven by the stock level falling below a threshold. They also experimented with different amounts of information sharing between organizations, and found that in the divergent assembly supply chains that they modeled, an assembly-to-order strategy, coupled with the sharing of both supply and demand information between organizations along the supply chain, was the most efficient. They also pointed out that their results reinforce the general point that information can substitute for inventory: If one has good information, uncertainty about demand can be reduced, and the required inventory level to satisfy orders can also be reduced as a consequence.

1.2.6   Electricity Markets

In many developed countries, in recent years, the electricity supply has been privatized. It is now common for there to be two or three electricity utilities that sell power to a number of distributors that in turn sell the electricity to commercial and domestic users. The change from a monopoly state-owned or state-regulated supplier to one in which there are a number of supply firms bidding into a market has inspired a range of agent-based models that aim to anticipate the effect of market regulations; changes to the number and type of suppliers and purchasers; and policy changes intended, for example, to reduce the chance of blackouts or decrease the environmental impact of generation (Bagnall & Smith, 2005; Batten & Grozev, 2006; Bunn & Oliveira, 2001; Koesrindartoto, Sun, & Tesfatsion, 2005; North, 2001).

In these models, the agents are the supply companies that make offers to the simulated market to provide a certain quantity of electricity at a certain price for a period, such as a day or an hour. This is also how the real electricity markets work: Companies make offers to supply and the best offer is accepted (different markets have different rules about what is meant by the "best" offer). Usually, the demand varies continuously, so supply companies have a difficult job setting a price for the electricity that maximizes their profit. A further complication is that the cost of generation can be very nonlinear: Matching peak demand may mean starting up a power station that is used for only a few hours.

By running the simulation, one can study the conditions under which the market price comes down to near the marginal cost of generation; the effect of mergers that reduce the number of supply companies; and the consequences of having different types of market "design," such as allowing futures trading. Most of the current models allow the agents to "learn" trading strategies using a technique known as *reinforcement learning* (see Section 5.2.1). A supply agent starts by making a bid using a pricing strategy selected at random from a set common to all the suppliers. If the bid is accepted and profitable, the value of this strategy is reinforced and the probability of using it again in similar circumstances is increased, or if it is unsuccessful, the chance of using it again is decreased (Roth & Erev, 1995).

1.2.7 Participative and Companion Modeling

Agent-based models have been used with success in rural areas in Third World countries to help with the management of scarce natural resources such as water for irrigation. This surprising use of agent-based models is due to their fit with "participative" (or participatory) research methods. As well as being used for research, multiagent models have been used as a support for negotiation and decision making and for training with, for example, the farmers in Senegal (D'Aquino, Le Page, Bousquet, & Bah, 2003), foresters and farmers in the Central Massif in France (Etienne, 2003; Etienne, Le Page, & Cohen, 2003), and the inhabitants of an atoll in Kiribati in the South Pacific (Dray et al., 2006).

The approach, also called *companion modeling* (Barreteau, 2003; Barreteau, Bousquet, & Attonaty, 2001; Barreteau, Le Page, & D'Aquino, 2003) involves building a multiagent system in close association with informants selected from the people on the ground. As a preliminary, the informants may be interviewed about their understanding of the situation, and they then engage in a role-playing game. Eventually, when sufficient knowledge has been gained, a computer model is created and used with the participants as a training aid or as a negotiation support, allowing the answering of "what-if" questions about possible decisions.

For example, Barreteau et al. (2001) describe the use of participative modeling in order to understand why an irrigation scheme in the Senegal River Valley had produced disappointing results. They developed both a role-playing game (RPG) and a multiagent system called SHADOC to represent the interactions between the various stakeholders involved in making decisions about the allocation of water to arable plots in the irrigated area. In this instance, the multiagent model was developed first and then its main elements converted to an RPG (in which the players were the equivalent of the agents in the multiagent model), partly to validate the agent-based model, and partly because the RPG is easier to use in a rural environment. The authors sum up the value of this approach as "enhancing discussion among game participants" and enabling "the problems encountered in the field and known by each individual separately [to be] turned into common knowledge."

1.3 The Features of Agent-Based Modeling

These examples, chosen to illustrate the spectrum of agent-based modeling now being undertaken, also provide examples of some characteristic features of agent-based modeling (Fagiolo, Windrum, & Moneta, 2006).

1.3.1 Ontological Correspondence

There can be a direct correspondence between the computational agents in the model and real-world actors, which makes it easier to design the model and interpret its outcome than would be the case with, for example, an equation-based model. For instance, a model of a commercial organization can include agents representing the employees, the customers, the suppliers, and any other significant actors. In each case, the model might include an agent standing for the whole class (e.g., "employees"), or it might have a separate agent for each employee, depending on how important the differences between employees are. The models of electricity markets described above have agents for each of the main players in the market.

1.3.2 Heterogeneous Agents

Generally speaking, theories in economics and organization science make the simplifying assumption that all actors are identical or similar in most important respects. They deal, for example, with the "typical firm," or the economically rational decision maker. Actors may differ in their preferences, but it is unusual to have agents that follow different rules of behavior, and when this is allowed, there may be only a small number of sets of such actors, each with its own behavior. This is for the good reason that unless agents are homogeneous, analytical solutions are very difficult or impossible to find. A computational model avoids this limitation: Each agent can operate according to its own preferences or even according to its own rules of action. An example is found in the models of supply chains, in which each business can have its own strategy for controlling inventory.

1.3.3 Representation of the Environment

It is possible to represent the "environment" in which actors are acting directly in an agent-based model. This may include physical aspects (e.g., physical barriers and geographical hurdles that agents have to overcome), the effects of other agents in the surrounding locality, and the influence of factors such as crowding and resource depletion. For example, Gimblett (2002) and colleagues have modeled the movement of backpackers in the Sierra Nevada Mountains in California to examine the effect of management policies in helping to maintain this area of wilderness. The agents simulated trekking in a landscape linked to a *geographical information system* that modeled the topology of the area. The environment also plays an important role in the models of industrial districts mentioned in the previous section.

1.3.4 Agent Interactions

An important benefit of agent-based modeling is that interactions between agents can be simulated. At the simplest, these interactions can consist of the transfer of data from one agent to another, typically another agent located close by in the simulated environment. Where appropriate, the interaction can be much more complicated, involving the passing of messages composed in some language, with one agent constructing an "utterance" and the other interpreting it (and not necessarily deriving the same meaning from the utterance as the speaker intended). The opinion dynamics models (Section 1.2.2) are a good example of the importance of interactions in agent-based models.

1.3.5 Bounded Rationality

Many models implicitly assume that the individuals whom they model are rational, that is, that they act according to some reasonable set of rules to optimize their utility or welfare. (The alternative is to model agents as acting randomly or irrationally, in a way that will not optimize their welfare. Both have a place in some models.) Some economists, especially those using rational choice theory, have been accused of assuming that individuals are "hyperrational," that is, that people engage in long chains of complex reasoning in order to select optimal courses of action, or even that people are capable of following chains of logic that extend indefinitely. Herbert Simon (1957), among others, criticized this as unrealistic and proposed that people should be modeled as *boundedly rational*, that is, as limited in their cognitive abilities and thus in the degree to which they are able to optimize their utility (Kahneman, 2003). Agent-based modeling makes it easy to create boundedly rational agents. In fact, the challenge is usually not to limit the rationality of agents but to extend their intelligence to the point where they could make decisions of the same sophistication as is commonplace among people.[1] Models of stock markets and the segregation model introduced in Section 1.2.1 are examples where agents have been designed with strictly limited rationality.

1.3.6 Learning

Agent-based models are able to simulate learning at both the individual and population levels. For example, the firms in the model of innovation networks described above (Section 1.2.4) are able to learn how to produce a more salable and more profitable product, and the sector as a whole (that is, all the agents in the model as a collection) learn over time which products will form a compatible set so that products from one firm will provide the components bought by another firm. Learning can be modeling in any or all of three ways: as individual learning in which agents learn from their own experience; as evolutionary learning, in which the population of agents learns because some agents "die" and are replaced by better agents, leading to improvements in the population average; and social learning, in which some agents imitate or are taught by other agents, leading to the sharing of experience gathered individually but distributed over the whole population (Gilbert et al., 2006). The model of innovation networks summarized above is an example of a model incorporating learning: The individual innovating firms learn how to make better products, and because poorly performing firms become bankrupt to be replaced by better start-ups, the sector as a whole can learn to improve its performance.

Some techniques for designing models that incorporate learning will be discussed in Section 5.2.

1.4   Other Related Modeling Approaches

The previous section has reviewed some areas where agent-based models have been useful. However, agent-based models are not appropriate for every modeling task. Before starting a new project, it is worth considering the alternatives. This section introduces two styles of modeling used in the social sciences that stand comparison with agent-based modeling: microsimulation and system dynamics.

1.4.1 Microsimulation

Microsimulation starts with a large database describing a sample of individuals, households, or organizations and then uses rules to update the sample members as though time was advancing. For example, the database might be derived from a representative national survey of households and include data on variables such as household members' age, sex, education level, income, employment, and pension arrangements. These data would relate to the specific time period when the survey was carried out. Microsimulation allows one to ask what the sample would be like in the future. For example, one might want to know how many in the sample would be retired in 5 years' time and how this would affect the distribution of income. If we have some rules about the likely changes in individual circumstances during the course of a year, these rules can be applied to every person in the sample to find what might have changed by the end of the first year after the survey. Then the same rules can be reapplied to yield the state of the sample after 2 years, and so on. After this aging process has been carried out, aggregate statistics can be calculated for the sample as a whole (for example, the mean and variance of the distribution of income, which can be compared with the distribution at the time of the survey) and inferences made about what changes are to be expected in the population from which the sample was drawn (Gupta & Kapur, 2000; Harding, 1996; Mitton, Sutherland, & Weeks, 2000; Orcutt, Merz, & Quinke, 1986; Redmond, Sutherland, & Wilson, 1998).

Microsimulation has been used to assess the distributional implications of changes in social security, personal tax, and pensions. For example, it can be helpful in evaluating the effects of changing the income threshold below which state benefits become payable (Brown & Harding, 2002). Experimental prototypes have also been developed in which there are several databases, describing not only individuals but also firms, and in which the aging process is affected not only by individual characteristics but also by macroeconomic variables such as inflation and the growth in gross domestic product (GDP); see the bibliography at http://www.microsimulation.org/.

An advantage of microsimulation models is that they start not from some hypothetical or randomly created set of agents but from a real sample, as described by a sample survey. Hence, it is relatively easy, in comparison with agent-based models, to read back from the results of the microsimulation to make predictions about the future state of a real population. There are two main disadvantages. First, the aging process requires very detailed transition matrices that specify the probability that an agent currently in some state will change to some other state in the following year. For example, one needs to know the probability that someone currently in employment will become unemployed 1 year later. Moreover, because this transition probability will almost certainly differ between men and women, women with and without children, young and old people, and so on, one needs not a single probability value but a matrix of conditional probabilities, one for each combination of individual circumstances. Obtaining reliable estimates of such transition matrices can be very difficult, requiring estimation from very large amounts of data. Second, each agent is aged individually and treated as though it is isolated in the world. Microsimulation does not allow for any interaction between agents and typically has no notion of space or geography. So, for instance, it is hard to take account of the finding that the chances of getting a job if one is unemployed are lower if one lives in an area where the unemployment rate is high.

1.4.2 System Dynamics

In the system dynamics approach to modeling, one creates a model that expresses the temporal cause-and-effect relationships between variables, but agents are not represented directly. One of the earliest and best known examples is Forrester's model of the world, which was used to make predictions about future population levels, growing pollution, and rates of consumption of natural resources (Forrester, 1971). System dynamics, as its name implies, models systems of interacting variables and is able to handle direct causal links, such as a growth in population leading to increased depletion of resources, and feedback loops, as when population growth depends on the food supply, but food supply depends on the level of the population (Sterman, 2000).

It is often convenient to represent a system dynamics model with a diagram in which arrows represent the causal links between variables. Figure 1.2 shows a typical, although simple, model of an ecosystem in which sheep breed in proportion to their population, wolves eat the sheep, but if there are too few sheep, the wolves starve. The rectangular boxes represent the stocks of sheep and wolves, the tap-like symbols are flows into and out of the stocks, and the diamond shapes are variables that control the rate of flow. The population of sheep increases as sheep are born, and the rate at which this happens is determined by the constant sheep-birth-rate. The diagram shows that sheep die at a rate that is a function of the number of sheep living (the curved arrow from the stock of sheep to the flow control labeled sheep-deaths), the probability that a wolf will catch a sheep (the arrow from the predation-rate variable), and the number of wolves (the arrow from the stock of wolves). Although this illustrative model is concerned with somewhat imaginary wolves and sheep, similar models can be constructed for

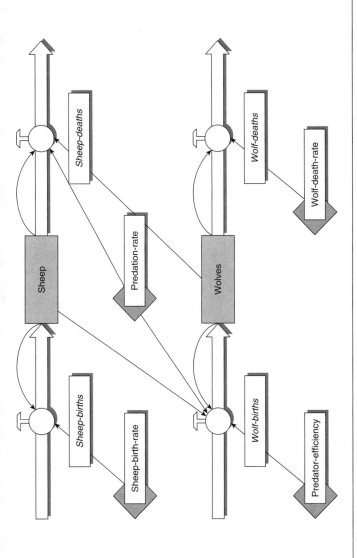

Figure 1.2 A System Dynamics Model of a Simple Ecosystem, With Wolves Eating Sheep According to the Lotka-Volterra Equations

SOURCE: Wilensky, U. (2005). NetLogo Wolf Sheep Predation (System Dynamics) model. http://ccl.northwestern.edu/netlogo/models/WolfSheepPredation (System Dynamics). Center for Connected Learning and Computer-Based Modeling, Northwestern University, Evanston, IL.

topics of sociological interest, such as the number of illegal drug users and enforcement agents, and public health epidemics (Homer & Hirsch, 2006; Jacobsen, Bronson, & Operations Research Society of America, 1985).

System dynamics is based on the evaluation of sets of simultaneous differential or difference equations, each of which calculates the value of a variable at the next time step given the values of other, causal variables at the current time step. Software such as Stella (http://www.iseesystems .com/) and NetLogo (http://ccl.northwestern.edu/netlogo/, described in more detail in Chapter 3) can help with drawing the diagrams and also execute the simulation by computing these equations.

In comparison with agent-based modeling, the system dynamics approach deals with an aggregate, rather than with individual agents. For example, in the wolves and sheep model, the simulation will compute the total population of sheep at each time step, but each individual sheep is not represented. This makes it hard to model heterogeneity among the agents; although one could, in principle, have a distinct stock for each different type of agent (e.g., a stock of white sheep, a stock of black sheep, a stock of mottled sheep, and so on), in practice this becomes extremely cumbersome with more than a few different types. It is also hard to represent agent behaviors that depend on the agent's past experience, memory, or learning in a system dynamics model. On the other hand, because they deal with aggregates, the system dynamics approach is good for topics where there are large populations of behaviorally similar agents. Thus, system dynamics was an appropriate method for Forrester's models of the global economy because individual action was unimportant and the focus was on the state of the world as a whole.

Note

1. Nevertheless, it has been found in several contexts, such as in modeling stock markets, that the aggregate behavior of agents with very little rationality (or "zero intelligence") matches the observed behavior at the macro level surprisingly well (Chan, LeBaron, Lo, & Poggio, 1999; Farmer, Patelli, & Zovko, 2005).

2. AGENTS, ENVIRONMENTS, AND TIMESCALES

We noted in the previous chapter that an agent-based model consists of a set of agents acting within an environment. In this chapter, we shall begin to show the principles of how agent-based models can be created and programmed.

2.1 Agents

Agents are conventionally described as having four important features (Wooldridge & Jennings, 1995):

1. *Autonomy.* There is no global controller dictating what an agent does; it does whatever it is programmed to do in its current situation.

2. *Social Ability.* It is able to interact with other agents.

3. *Reactivity.* It is able to react appropriately to stimuli coming from its environment.

4. *Proactivity.* It has a goal or goals that it pursues on its own initiative.

However, these characteristics are not very helpful when designing an agent, and it requires a lot of imagination to see the agents of most models possessing all these features. A more helpful way of describing agents is to say that they have the following characteristics:

1. *Perception.* They can perceive their environment, possibly including the presence of other agents in their vicinity. In programming terms, this means that agents have some means of determining what objects and agents are located in their neighborhood.

2. *Performance.* They have a set of behaviors that they are capable of performing. Often, these include the following:

    a. *Motion.* They can move within a space (the environment).
    b. *Communication.* They can send messages to and receive messages from other agents.
    c. *Action.* They can interact with the environment, for example, picking up "food."

3. *Memory.* They have a memory, which records their perceptions of their previous states and actions.

4. *Policy.* They have a set of rules, heuristics, or strategies that determines, given their present situation and their history, what behaviors they will now carry out (Conte & Castelfranchi, 1995).

These features can be implemented in many different ways. In this section, we describe three common approaches: building a simulation and its agents using a programming language directly, using a production rule system, and using neural networks.

2.1.1 Ad Hoc Programming

Almost all agent-based models are built using an object-oriented programming language, such as Java, C++, or even Visual Basic. This book cannot teach you object-oriented programming (there are many good introductions, including Eckel, 2005, and Niemeyer & Knudsen, 2005), but because the idea of object-oriented programming is so important to agent-based modeling, here is a short introduction to its main features.

Object-oriented programming develops programs as collections of objects, each with its own set of things it can do. An object is able to store data in its own *attributes*, can send *messages* to other objects, and has *methods* that determine how it is able to process data. The general programming advantage of object orientation is that it provides a high level of modularity: For example, the details of how an object's methods work can be changed without upsetting the rest of the program. An additional advantage for agent-based modeling is that there is an affinity between the idea of an agent and an object: It is natural to program each agent as an object.

There are several basic concepts in object-oriented programming:

• A *class* is an abstract specification of an object, including its attributes and its methods. For example, a program might include a class called "Company" to represent a firm in a model of an economy. A Company might have attributes such as its capitalization, its number of employees, and the type of product it sells. It might also have methods that describe the processes involved in selling the Company's products to customers and buying the Company's materials from suppliers, which could be other companies in the model. Classes may be specialized to form more specific classes. For example, the general Company class could be specialized to yield Manufacturing companies and Distribution companies. Each specialized class inherits the attributes and methods of its more abstract class, and may add new attributes and methods or override the ones it inherits. For example, the Manufacturing class would need methods that determine the volume of product that can be made from unit volumes of materials; the Distribution class would not need this method.

- As the program runs, classes are *instantiated* to form objects. For example, the Manufacturing class might be instantiated to yield two objects, one representing XYZ, Inc. and the other ABC, Inc. Although the two objects have the same methods and the same attributes, the values of their attributes (e.g., their capitalization, size, and type of product) can differ. An object's methods may send messages to another object, thus affecting its state. For example, one of the methods of the XYZ object may send a message to the ABC object requesting it to sell some of its products; a method of the ABC object might respond with the number and price of the products it wishes to sell.

As this brief summary should suggest, it is a short step from object orientation to agent-based modeling. One creates a class for each type of agent, provides attributes that retain the agents' current state (*memory*), and adds suitable methods that observe the agents' environment (*perception*) and carry out agent actions (*performance*) according to some rules (*policy*). In addition, one needs to program a scheduler that instantiates the required number of agents at the beginning of the simulation and gives each of them a turn to act. Chapter 3 includes an example of building such a simulation.

2.1.2 Production Rule Systems

As we noted in the previous section, the agents in most models need to have the ability to perceive the state of their environment, receive messages from other agents, proactively select behavior to perform depending on their current state, and send messages to other agents. One way to achieve these is to endow the agents with the following:

- *A Set of Rules of Behavior.* These rules determine what the agent will do. One or more rules from the set will be selected depending on the current state of the agent. Such rules are often called "condition-action" rules because they include both a condition component (what must be true if the rule is to be used) and an action component (what will be done to carry out the rule). For example, one such rule might be "If I can see food in the vicinity, then I will move a step toward it."
- *A Working Memory.* This will consist of variables that store the agent's current state. For example, the working memory might store the agent's current location and its current energy reserve.
- *A Rule Interpreter.* This is some program code that uses the working memory in order to select which rule should be activated, or "fired." It may need to handle the situation where the condition component of more than one rule is satisfied, and therefore some means of choosing between the rules is needed.

- *An Input Process.* This will collect messages and perceptions from the environment and store them in working memory for processing by the rules.
- *An Output Process.* This will pass messages to the environment, en route to the agents that are the intended recipients.

Such an arrangement is called a *production system*, and the rules are *production rules* (Nilsson, 1998; Waterman & Hayes-Roth, 1978). The correspondence between the elements of a production system and the desirable features of an agent listed above should be clear. A relatively simple production system can be constructed from a toolkit such as JESS (the Java Expert System Shell, http://www.jessrules.com/) (Friedman-Hill, 2003). Alternatively, there are a few much more elaborate systems that are based on psychologically plausible models of human cognition, such as Soar (Laird, Newell, & Rosenbloom, 1987; Wray & Jones, 2006; Ye & Carley, 1995), CLARION (Sun, 2006), and ACT-R (Taatgen, Lebiere, & Anderson, 2006).

2.1.3   Neural Networks

Another way of providing the features required for an agent is to use *artificial neural networks* (ANNs). Their design is loosely based on biological neurons such as those in the brain that are capable of learning or pattern matching. ANNs can be constructed in hardware, but more usually they are software programs or libraries. An ANN consists of several (usually between 10 and 100) *units* arranged in stacked layers (Engelbrecht, 2002; Garson, 1998). A unit is connected to all the units in the layer behind it and in front of it (Figure 2.1). Each connection conveys a signal (in a software implementation, a number representing the magnitude of the signal) from a unit in one layer to a unit in the next. A unit takes in all the signals arriving from the preceding layer, multiplies each by a *connection weight*, sums the products, applies some nonlinear mathematical transformation (e.g., the logistic or the hyperbolic tangent) to the sum, and then sends that signal on to all the units in the next layer. The first layer of units receives inputs, one per unit, from outside the network, and the last layer delivers an output consisting of a signal from each unit in the layer.

In the simplest application of an ANN, the network is trained to recognize patterns, outputting a specific output signal every time a particular pattern is applied to the inputs. For example, an ANN might be trained to recognize bar codes and output the digits corresponding to each pattern of bar code lines. Before training, all the connection weights are set to random values. Applying a bar code as input to the first layer results only in a random pattern of signals at the output layer. A method called the *back-propagation algorithm* is then applied. This compares the actual output with the desired

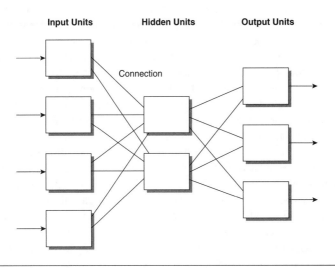

Input Units   **Hidden Units**   **Output Units**

Connection

Figure 2.1   Schematic Diagram of the Units in an Artificial
Neural Network

output (e.g., it compares the random digits produced by the untrained network with the string of digits the bar code is supposed to generate) and modifies the connection weights to bring the output slightly closer to the desired output. Then another input (e.g., another bar code) is applied, the actual output is compared with the desired output, the connection weights are modified again, and this process of training continues until the ANN performs well enough at its recognition task. This can involve tens of thousands of training examples, but once the network is properly trained, it can be used to decode inputs that it has not seen before, or it can be applied to model a theory of social groups or financial markets (e.g., Beltratti, Margarita, & Terna, 1996; Kitts, Macy, & Flache, 1999).

Another option, one that has been used in agent-based models (Acerbi & Parisi, 2006; Cecconi & Parisi, 1998), is to create a large population of ANNs (each corresponding to one agent). The agents are all initialized with a random set of connection weights and are set a task such as gathering "food" from a landscape. An agent's perception of whether there is food in front of it is fed into the ANN inputs, and the outputs are linked to the agent's actions, such as move and eat. The agent is given an initial quantity of energy, some of which is used on every time step. If the energy declines to zero, the agent "dies" and is removed from the simulation. However, an agent can boost its energy by eating food, which is scattered around the landscape. Because of the random connection weights with which the agent's ANN is initialized, most agents will not succeed in finding and eating food and will quickly die, although some will succeed. These more successful agents reproduce, giving their offspring similar connection weights as their own, but with slight mutations. Gradually, the population of agents will learn food harvesting behavior. Although some variation will remain, the ability of the average agent will improve, generation by generation. Thus, instead of directly changing connection weights as with the backpropagation approach, these models use a form of Darwinian selection to breed agents with better adapted weights.

In comparison with programming agents using production rules, neural networks have the advantage that they are more flexible and can develop behaviors that the researcher had not considered. A neural network could also be considered to be more akin to the human brain than a production rule system (although it should be recognized that an artificial neural network is very different in the way that it works from the network of biological neurons in the brain, as well as being much simpler). On the other hand, once a neural network has been trained, it is almost impossible to discover why it works: All one has is a matrix of connection weights that do not provide a comprehensible explanation for its behavior in terms of the activity of the units. With a production system, one can point to the rules to explain the agent's behavior. Finally, models using neural networks that are trained through an evolutionary process are very demanding of computing resources. In order to evolve a functioning agent, one needs a large initial population and many generations of development.

2.2   Environments

The previous section has described three ways of building agents. The agents act within an environment, which provides the channel of communication between agents and may also include nonreactive objects, such as obstacles or energy sources.

It is convenient to route all communication between agents through the environment, not only because this is the "natural" way to do it, corresponding to the role of the environment in human affairs, but also because it makes monitoring the agents easier. It also means that messages from one agent to another can be temporarily stored in the environment (*buffered*), reducing the likelihood that the results of the simulation will depend on the accidents of the order in which agent code is executed. When buffering, messages from agents are stored in a temporary variable until all the agents in the simulation have had their turn. Then, these stored messages are delivered to the recipients.

In many models, the environment will include passive objects, such as landscape barriers, "roads" or links down which agents may travel, resources to provide agents with energy or food, and so on. These can be programmed in much the same way as agents, but more simply, because they do not need any capacity to react to their surroundings.

Because the environment is often straightforward to program, it tends to get neglected; yet environmental influences are generally very important in the real world. Much of the complexity of human life comes from dealing with a complex environment. We also often use the environment as a memory (e.g., placing objects in particular locations to remind us that action needs to be taken), as a store of value (e.g., money and other forms of wealth), and as a technological aid to make some actions easier (e.g., devices that provide communication and transport services). However, it is not often that researchers recognize this environmental complexity in designing simulation models.

2.3 Randomness

One way of abstracting from the complexity of the real world when designing a model is to build in some randomness. For example, in a model of an industrial network (Section 1.2.4), one might want to abstract from the firms and interfirm linkages to be found in a particular industrial sector in order to develop a general model of networks between firms. However, the question is, then, which linkages should be built into the model? One answer is to choose pairs of firms at random and create links between these pairs. This may be done in several ways, depending on the intended distribution of linkages. For example, the chance of a firm being linked to another could be uniform for all firms (a "random network"), could be related to the number of firms already linked to the other firm ("preferential attachment") (Barabási, 2003; Barabási & Albert, 1999), or could be arranged so that the links form a local cluster of strongly connected firms with a few long-distance connections between clusters (a "small-world" network) (Watts, 1999; Watts & Strogatz, 1998). These alternative schemes provide networks with different structural properties that may have quite different behaviors (Pujol, Flache, Delgado, & Sangüesa, 2005).

Randomness may also be used to model communication errors and the influence of "noise." For example, Axelrod (1997b; Axelrod & Dawkins, 1990) developed an influential model of the dissemination of culture (Axelrod, 1997c) to explain why, if the beliefs and attitudes of people who interact tend to become more alike, all differences do not eventually disappear. His model displays the emergence of stable regions of homogeneous cultural traits, such as dialects, nationalistic beliefs, or religious customs. In his model, agents have "tags," a set of five numbers that describe their own cultural traits on five cultural features or dimensions.[1] The chances of two agents interacting depends on the similarity of their tags. If they do interact, one feature from the five is chosen at random, and one agent adopts the other agent's value for that feature. After several thousand interactions, distinct regions emerge in which all the agents share the same traits and have no traits in common with agents in other regions. Axelrod comments that his model shows how local convergence can give rise to global polarization. However, later work (Klemm, Eguíluz, Toral, & San Miguel, 2003) showed that this behavior is critically dependent on the agents' copying the value of the other agents' cultural traits exactly and there also being no "cultural drift" in an agent's traits. If, instead, an agent adopts a slightly distorted copy of the other agent's traits, or the agent's beliefs occasionally change randomly, a monoculture may emerge, rather than there being several distinct regions, each with different cultures. This is an example of how random variation or "noise" can make a radical difference to the outcome of a model.

2.4   Time

In most cases, simulations proceed as though orchestrated by a clock, beat by beat. At each beat, all the agents are given a turn. Thus, time is modeled in discrete time steps. Each time step lasts for the same simulated duration. The simulation starts at time step zero and proceeds as long as necessary, or until all the agents are "dead" or out of action.

There are three issues about time that need consideration when designing agent-based models using discrete time steps:

1. *Synchronicity.* We have already mentioned in the previous section that one needs to be careful about the timing of messages sent from one agent to another. For example, if Agent A sends a message to Agent B, and B replies, and then C sends a message to B, the outcome might be quite different from the results of A sending a message to B, then C sending a message to B, and then B sending a message to A (consider these sequences in the context of a model of "insider trading," for example). This is an example of a more general issue of the order of agent invocation (Huberman & Glance, 1993).

With an ordinary computer, because only one thing can happen at once, agents cannot, in fact, engage in action simultaneously. The three possible ways to work around this are as follows:

   a. Invoke each agent in sequential order (Agent 1, Agent 2, Agent 3, Agent 4, Agent 1, Agent 2, Agent 3, Agent 4, . . .) (*sequential asynchronous execution*). This is rarely a good solution, because the performance of the model may be greatly influenced by the order that is used.

b. Invoke each agent in a different random order at each time step (*random asynchronous execution*). The advantage of this is that the effect of the ordering can be investigated by running the simulation several times, with a different ordering each time.

c. Invoke each agent in any convenient order, but buffer all interactions with the agents' environment so that all inputs to agents are completed before all outputs (*simulated synchronous execution*). This is the best option if it can be achieved, although it can be complicated to arrange, and sometimes the requirements of the model prevent it.

2. *Event-Driven Simulation*. These three ways of arranging the ordering of agent invocation assume that all the agents need to have a chance for action in each time step, although some agents may actually do nothing during their slot. A different approach is possible: to use an event-driven design in which only those agents that need to take action are invoked. The idea of event-driven simulation is that, instead of having a time step of constant duration, the simulation skips from one event to another. The simulation "clock" is wound forward until the time of occurrence of the next event. For instance, suppose that we are designing a simulation of organizational decision making and have a model in which the agents are decision-making committees. The focus of the model is on the decisions that each committee makes and how these decisions are passed from one committee to another. What happens between committee meetings is of no concern, and the meetings themselves can be considered to be instantaneous. In such a model, having a regular time step would be inefficient, because nothing of interest would be happening at most steps. Instead, the model could be designed to "jump" from the time of one committee meeting to the next.

3. *Calibrating Time*. With both the regular and event-driven modes of simulation, there is often a problem of matching the simulation time with real time. For example, if one has a model of consumer behavior in which one wants to study the reactions of consumers to the introduction of a new product, a matter of some interest will be the time it takes for a majority of consumers to adopt the product. The model might indicate how many time steps this takes, but how does one translate this into weeks or months of real time? A solution is to observe the process in reality and match the shape of the adoption trend against the output from the simulation, but this will give only approximate answers. Moreover, in this and other examples, there remains a difficulty about what to take as the start or zero time points in the simulated and real worlds. Although a simulation starts at a well-defined moment, in the real world, it is rare that any activity (for example,

the marketing of a new product) commences at one clear moment in time. These are issues that one needs to watch out for; there is no general solution that always applies.

Note

1. A number of subsequent papers by different authors have used such tags to explore the emergence of differences between groups of agents (e.g. Edmonds, 2006; Hales, 2000, 2002; Riolo, Cohen, & Axelrod, 2001).

3. USING AGENT-BASED MODELS IN SOCIAL SCIENCE RESEARCH

Over the past decade, agent-based modeling has developed a more or less standardized research process, consisting of a sequence of steps. Like most social science methods, this is an idealization of the procedures actually carried out, and in practice, several of the steps occur in parallel and the whole process is performed iteratively as ideas are refined and developed. Nevertheless, it is useful to have these steps made explicit as a guide to the conduct of agent-based modeling research.

At an early stage in the research, it is essential to narrow down the general research topic to some specific research question. A *research question* is something that the work should have a realistic chance of answering. If the question is too vague or too general, it will not be much use, and the research will be disappointing as it will not be able to provide the hoped-for answers. It is always better to err on the side of specificity: Be too focused rather than too ambitious. It is sometimes helpful to think of defining the research question as like stripping away the layers of an onion, from the general area of investigation, through the particular topic, to a question that could be answered in no more than a brief statement of what you have discovered.

As we have noted above, the usual kind of research question that agent-based models are used to study are ones where regularities at the societal or macro level have been observed, and the issue is how these may be explained. Economists often call these regularities *stylized facts* (Kaldor, 1961). For example, the Schelling model described in Chapter 1 starts with the observation that neighborhoods are ethnically segregated and seeks to explain this through modeling individual household decisions. The electricity market models also described in Chapter 1 aim to explain (and predict) patterns of electricity supply and market pricing in terms of the motivations of suppliers.

After having specified the research question clearly and identified the macro-level regularities that are to be explained, the next step is to specify the agents that are to be involved in the model. They may be all of one type, or there may be different types. For example, while the models of opinion dynamics reviewed in Chapter 1 involve only one type of agent, the individuals whose opinion changes are being simulated, some of the industrial districts models mentioned in Chapter 1 involved several distinct types of firms. For each type of agent, one needs to lay out the agent's behavior in different circumstances, often as a set of condition-action rules (see Section 2.1.2). It is helpful to do this in the form of two lists: one that shows all the different ways in which the environment (including other agents) can affect the agent, and one showing all the ways in which the agent can affect the environment (again, including other agents). Then, one can write down the conditions when the agent has to react to environmental changes, and the conditions when the agent will need to act on the environment. These lists can then be refined to create agent rules that show how agents need to act and react to environment stimuli.

At this stage, one will have a good idea of the types of agents and their behaviors that are needed in the model. It will also be necessary to consider what form the environment should take (for instance, does it need to be spatial, with agents having a definite location, or should the agents be linked in a network?) and what features of the model need to be displayed in order to show that it is reproducing the macro-level regularities as hoped. Once all this has been thought through, one can start to design and develop the program code that will form the simulation.

After the model has been constructed, one begins the long process of checking that it is correct. Informally, this is called debugging; more formally, it is *verification*. Verification is the task of ensuring that a model satisfies the specification of what it is intended to do. It is quite different from *validation*, which is checking that the model is a good model of the phenomenon being simulated. One can have a simulation that satisfies the verification criterion, because it runs as it is supposed to do, but if the specification is a poor description of the target in the social world, it is not a valid model.

Following successful verification, one can embark on validation. The primary criterion of validation is whether the model shows the macro-level regularities that the research is seeking to explain. If it does, this begins to be evidence that the interactions and behaviors programmed into the agents explain why the regularities appear. However, one must guard against alternative explanations. There may be other, equally or more plausible agent behaviors that lead to the same macro-level regularities. Therefore, one needs to engage in a *sensitivity analysis* to see whether, when model parameters are changed, the outcomes alter, too. It is also important to consider whether quite different agent behaviors could lead to the same results, in particular whether a simpler model leads to the same conclusions (if it does, the simpler model should normally be preferred to the more complicated one, using the principle that simple explanations are better than complicated ones if both are equally good at explaining).

Having thus explored the macro behavior of the model, it is then desirable to compare the output of the model with empirical data from the social world. As we shall see later, such comparisons between model outputs and data are not easy to carry out and often do not lead to the clear answers that one might expect. Most models are stochastic, that is, involve random processes, so one does not know whether any difference between the model output and observed data is due to random chance or a bad model. There are also often considerable difficulties in collecting valid and reliable data, especially the data observed over long periods of time that one needs to compare with model outputs.

Finally, one can draw some conclusions, hopefully answering the research question that started the process. In addition, if one has confidence in the model, one can experiment with it, perhaps to identify regularities that had been previously unsuspected.

3.1   An Example of Developing an Agent-Based Model

In this section, the process of developing a simple agent-based model will be described using a simulation of "collectivities" (Gilbert, 2006) as an example. In Chapter 4, we shall see how this model could be programmed.

A number of related social phenomena are hard to model, or even to describe, because their boundaries are fluid, the people involved are constantly changing, and there is no single characteristic shared by all those involved. Examples include the following:

- Youth subcultures, such as "punks" (Widdicombe & Wooffitt, 1990) or "goths" (Hodkinson, 2002)
- Scientific research areas or specialties (Gilbert, 1997)
- Art movements such as the Pre-Raphaelites or the Vorticists (Mulkay & Turner, 1971)
- Neighborhoods, such as Notting Hill in London or the Bronx in New York (O'Sullivan & Macgill, 2005)
- Members of armed revolutionary or terrorist movements (Goolsby, 2006)
- Industrial sectors such as biotechnology (Ahrweiler, Pyka, & Gilbert, 2004)

Although one can easily point to familiar examples, and although they are very common and easily identified, it is difficult to put one's intuitions about them on a firmer footing. For a start, there is no commonly accepted word with which to name the phenomenon. The terms *subculture, area, neighborhood, specialty,* and *movement* are used for particular types, but none of these words is appropriate for describing all of them. A closely related concept is "figuration" (Elias, 1939/1969), although strictly this should be applied only to individuals, not to organizations or other types of actors. In this section, we use the term *collectivity* as the generic term, for lack of a better one. Note that the units making up the collectivity may be people (as in most of the examples above) or organizations (e.g., biotechnology firms).

A second barrier to gaining a better understanding of collectivities is that, by definition, there is no definite boundary around them. This means that it is impossible to count their members and therefore to engage in the more common kinds of quantitative analysis of their development over time, their incidence, and so on.

Third, the way in which collectivities arise from the actions of their members is not easily understood. It is the purpose of the model to be developed here to suggest how some plausible assumptions about individual action (*micro foundations*) could yield the collectivities that are observable at a macro level.

3.1.1 Macro-Level Regularities

In all collectivities, the following seem to hold, to a greater or lesser extent:

- Although instances of collectivities are usually easily named and described at the aggregate level, precise definitions can prove to be rather slippery and open to negotiation or argument (e.g., there are many slightly different areas that can be described as Notting Hill, from the official local government area to the locality within which the film of the same name was shot).

- There is no accepted consensual definition that can be used to sort those who are "in" from those who are "out" (or members from nonmembers). For example, whereas some might think that a person is a "punk" because of the way that he or she dresses, this assignment might be contested by others (including the person him- or herself) by pointing to the person's beliefs, behavior, or acquaintances, all of which could alternatively be relevant for making a decision on membership. In particular, there is no one observable feature that all those who are "in" and none of those who are "out" possess. Collectivities are not, for example, formal organizations, where being an employee with a written or verbal contract distinguishes those who are members; political parties, where, at a minimum, a formal declaration of support is required and defines membership; or social classes, where externally specified objective criteria are used to sort people (typically one's occupation).

- Nevertheless, many of the members will share characteristics in common (e.g., the scientists in a research area may have similar education, have carried out similar previous research, and be known to each other, even if there is no technique, theory, or object of research with which all of those without exception in the research area are involved).

- Membership of the collectivity entails possessing some related knowledge (e.g., the science of the specialty, or whatever is accepted as "cool" in a youth culture, or the local geography of Notting Hill). However, no member possesses all the knowledge: Knowledge is socially distributed.

- The features that are thought to be relevant to the collectivity change. For example, researchers do not continue to work on exactly the same problems indefinitely; once they have solved some, they move on to new ones, but still within the same research area. Most political movements change their manifestos over time to reflect their current thinking and the social problems that they see around them, although they remain the same movements, with many of the same adherents. Youth cultures are constantly changing the items that are considered to be in fashion.

- Some of the people involved are widely considered (e.g., by the others) as being more central, more influential, of greater status, or as leaders as compared with others. For example, some scientists are considered to be more eminent than others, some members of subcultures are more "cool" than the rest, and so on.

3.1.2 Micro-Level Behavior

One of the features common to collectivities mentioned in the previous section is that the actors (that is, the people or organizations that make up the collectivity) have some special knowledge or belief (e.g., for scientists, knowledge about their research area; for youth subcultures, knowledge about what is currently fashionable). Even though this knowledge is socially distributed among the members of the collectivity, so that not every member has the same knowledge, possession of it is often a major feature of the collectivity (Bourdieu, 1986). In the model, we assume that all individuals, members and nonmembers, have some knowledge, but what this knowledge is varies both between actors and over time. We use this knowledge to locate the actors: The position of the actor at a moment in time in an abstract knowledge space is a function of the knowledge that he or she possesses at that time.

A second assumption is that some actors are of higher status than others and that all actors are motivated to try to gain status by imitating high-status actors (by copying their knowledge). For example, in a collectivity driven by fashion, all actors will want to be as fashionable as they can, which means adopting the clothing styles, musical tastes, or whatever of those whom they perceive to be of the highest status (Simmel, 1907). However, status is also a function of rarity: An actor cannot remain of high status if there are many other actors with very similar knowledge. For example, a fashion icon must always be ahead of the hoi polloi; a scientist will be heavily cited only if his or her research is distinctive; a revolutionary will earn the respect of colleagues only if he or she stands out in comparison with the "foot soldiers."

Third, we assume that the highest status actors want to preserve this status, which they cannot do if they start to be crowded out by followers who have been attracted to them. In this situation, we assume that high-status actors are motivated to make innovations, that is, to search out nearby locations in knowledge space where there are not yet crowds.

There are thus two countervailing tendencies for actors—on one hand, they want to get close to the action; on the other, they want to be exclusive and can do so by changing the locations that represent the heights of status. As we shall see, working out this tension yields patterns at the macro level that are typical of collectivities.

3.1.3 Designing a Model

Related Models

There are several generic models that deal with similar issues:

1. *Boid models* (Reynolds, 1987) have agents that try to maintain a desired distance away from all other agents and thus appear to move with coordinated motion. Agents have three steering behaviors: separation, to avoid nearby agents; alignment, to move in the same direction as the average of nearby agents; and cohesion, to move toward the average position of nearby agents. The effect is that agents move as in a flock of sheep or a school of fish. These models illustrate the effect of having agents carrying out actions that are in "tension": The separation behavior is in tension with the cohesion behavior, for instance. However, there are no notions of seeking status or innovation in these models.

2. *Innovation models* (e.g., Ahrweiler & Gilbert, 1998) have agents that are able to learn and act according to their current knowledge. Agents also exchange knowledge and create new knowledge. However, there is no

specific idea of collectivity in these models. The set of agents involved in innovation is predetermined.

3. *The minority game* (Slanina, 2000) is one example from a large literature. This model, also called the El Farol Bar model, has agents who wish to go to the bar, but only when a minority of the other agents also choose to go there. The agents make a decision based on their own past experience of the number they previously encountered at the bar. Each agent has a number of strategies that he or she uses in combination with his or her memory of the outcome of recent trips to the bar to make a decision on whether to visit the bar at the current time step. The strategies are scored according to their success (whether, when the agent arrives at the bar, it is overcrowded or not), and unsuccessful strategies are dropped. Over time, a dynamic equilibrium can be established, with the number of agents at the bar matching the threshold that agents use to judge that there are too many agents there. This model has some features of the problem addressed herein, but there is no representation of a collectivity.

The Model

The collectivities model consists of a surface over which agents are able to move. The surface is a *toroid* with each point representing one particular body of knowledge or set of beliefs. The agents thus move, not in a representation of physical space, but rather in "knowledge space." Although it may be oversimplifying to represent a knowledge space in two dimensions (more exactly, on the surface of a toroid), it makes for easier visualization.

An agent's movement in the knowledge space represents its change in knowledge. Thus, if an agent imitates another agent, it would be viewed as moving toward that agent in the knowledge space, whereas if it innovates and discovers knowledge that other agents do not have, it would be viewed as moving away from other agents into previously empty areas of the space.

Agents are initially distributed at random on the surface. Agents have no memory of their own or other agents' previous positions. Each agent does the following:

1. It counts how many other agents there are in its immediate neighborhood.

2. If the number of agents is above a threshold, it turns to the direction opposite to the average direction of travel of other nearby agents and then moves a random distance.

3. If the number of agents is equal to or below the threshold, it looks around the locality to find a relatively full area and then moves a random distance from its present location in the direction of the center of that area.

Each agent acts asynchronously, repeating this sequence of actions indefinitely. There are four parameters required by this algorithm (see Figure 3.1):

1. The radius of the circular area surrounding an agent within which the number of agents is counted to determine whether the agent is "crowded" or "lonely" (*local-radius*)

2. The threshold number of agents below which the agent is "lonely" and above which the agent is "crowded" (*threshold*)

3. The radius of the circular area surrounding an agent in which the agent, if lonely, counts the number of agents to find where there is a maximum or, if crowded, finds the average direction of agent movement in order to determine the direction in which it is to move (*visible-radius*)

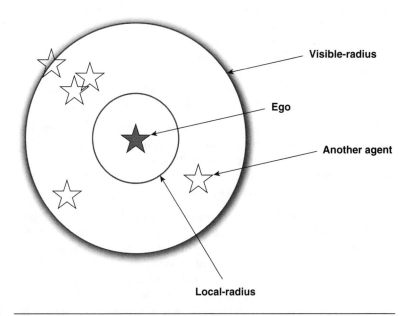

Figure 3.1   Parameters for the Collectivities Model

4. The distance that an agent moves; the distance is chosen randomly from a uniform distribution with this parameter as the maximum (*speed*)

3.1.4   Verification

The verification and validation of the collectivities model will be discussed in Chapter 4, after we have described how to program it. First, we shall consider some general issues that arise in verifying agent-based models.

3.2   Verification: Getting Rid of the Bugs

You should assume that, no matter how carefully you have designed and built your simulation, it will contain bugs (code that does something different from what you wanted and expected). When you first run a new simulation model, it is likely that it will have many bugs, most of them easily observable because the simulation gives anomalous results or crashes. More worrying, it is quite likely that even when you have worked on the code to remove the obvious bugs, some will still lurk to catch you unaware. As a rule of thumb, think of the number of bugs remaining as following a negative exponential—the number decreases rapidly at first, but then levels off and never hits zero. Even published simulations sometimes seem to suffer from bugs and misinterpretations (for examples, see Edmonds & Hales, 2003; Galan & Izquierdo, 2005; Rouchier, 2003).

There are techniques for reducing the chances of including bugs and for making finding them easier (Ramanath & Gilbert, 2004). Here are some:

• *Code Elegantly*. When you are writing the simulation program, do it carefully and steadily; don't rush to get the code working and take short-cuts. Any time saved will be lost in extra time needed for debugging. Using an object-oriented language and using variable names that are meaningful in the context of your model will help.

• *Include Lots of Output and Diagnostics*. It will be hard to find bugs in sections of code that give no output to show what is happening as the program runs. Don't be satisfied with only displaying the results of the simulation; at least during the debugging phase, you will need to display intermediate values also. Some care will be needed to decide what to display so that you get good diagnostics, but are not so overwhelmed that you cannot find the symptoms of the bugs because of the amount of other output in which they are buried.

• *Observe the Simulation, Step by Step.* Run the code one line or one function at a time, observing how the values of variables, parameters, and attributes change and checking that they alter in the expected way. Although this can be slow and tedious, it does help to ensure that the code is doing what was intended, at least for the runs that are observed. Often, programming environments provide features to make stepping through code easier to manage.

• *Add Assertions.* If you know that variables must take some values and not others, check for valid values as the simulation runs, and display a warning if the value is out of range (such checks are known as assertions). For example, if two agents cannot occupy the same spatial location, at each time step check and report if this requirement is violated.

• *Add a Debugging Switch.* You may worry that all the code needed to identify bugs, such as assertions and diagnostics, will slow down the simulation unacceptably. Include a global variable in your program that can be set to a debugging level: from none to maximum. Precede each debugging statement with a test of this variable, to see whether the statement should be run at the current debugging level (it is possible in some languages to achieve the same effect with conditional compilation).

• *Add Comments and Keep Them Up to Date.* All programming languages allow you to insert comments—text for programmers to read that is not executed as program code. Use this feature to add comments to every function, procedure, method, and object. The comments should describe what the following block of code does and how it does it, but at the conceptual level, not at the implementation level (that is, do not paraphrase the program code, but state what the code is intended to achieve). As a rule of thumb, there should be about one third as many lines of comment as there are lines of code. Comments can easily become obsolete, describing the program as it used to be, rather than as it is. Reserve some time to update the comments at regular intervals. In writing comments, assume that the reader is someone who can program as well you can, but who knows little or nothing about your model (after a couple of months away from the program, this may be a good description of you, so do not avoid writing comments with the excuse that no one else will see your code).

• *Use Unit Testing.* Unit testing is an increasingly popular software engineering technique for reducing bugs (Link & Fröhlich, 2003). It consists of writing some test code to exercise the program at the same time as you write the code itself. The idea is to develop the program in small, relatively self-contained pieces, or units. A test "harness" is created that will supply the unit with a sequence of inputs and check the results against a list of expected outputs. The test harness then automatically runs through each input, checking that the expected output is, in fact, generated. Once the unit has passed all of its tests, you can move on to writing the next unit. This may involve making some changes to the first unit, which must then be put through its test sequence again to ensure that the changes have not introduced any new bugs. When many units have been written, the test harness is used to automate performing all the tests on all the units, thus giving some guarantee that bugs have not inadvertently crept in as a result of developing the code. As the program develops, additional tests should be written to verify assemblies of several units and the interaction between them.

• *Test With Parameter Values for Known Scenarios.* If there are any scenarios for which the parameters and the output are known with some degree of certainty, test that the model reproduces the expected behavior. This is the test that most people carry out first of all on a model, but it is a rather weak test and, by itself, will not give much confidence that the simulation is free of bugs.

• *Use Corner Testing.* Test the model with parameter values that are at the extremes of what is possible and ensure that the outputs are reasonable (the name comes from the idea that such parameter values mark the corners of a parameter space enclosing all possible parameter values). For example, test to see what happens when your simulation is run without any agents and when it is run with the maximum number that your model allows.

3.3   Validation

Once one has developed an agent-based model, it seems obvious that one needs to check its validity, that is, whether it is, in fact, a good model of what it purports to represent. However, both the theory and practice of validation are more complicated, and more controversial, than one might at first expect. The issues are related to the various objectives aimed at by modelers, which imply different criteria for validation, and the sheer difficulty of acquiring suitable social science data in sufficient quantity to allow systematic validation (Troitzsch, 2004). We begin by considering the conceptual issues before discussing some techniques for carrying out validation.

Agent-based models can be directed primarily at formalizing a theory (for example, Schelling's residential segregation model; see Chapter 1), in which case the model is likely to be pitched at a very abstract level; or they can be aimed at describing a wide class of social phenomena, such as the development of industrial districts or the behavior of consumers; or they can be intended to provide a very specific model of a particular social situation, such as some of the models of electricity markets mentioned in

Chapter 1, where the precise characteristics of one market, including the location of power plants and pattern of consumer demand, are relevant. These types of agent-based models require rather different approaches to validation (Boero & Squazzoni, 2005).

3.3.1 Abstract Models

The aim of abstract models is to demonstrate some basic social process that may lie behind many areas of social life. A good example is Epstein and Axtell's pioneering book on *Growing Artificial Societies* (Epstein & Axtell, 1996), which presents a series of successively more complex models of the economics of an artificial society. Another example is the model of collectivities introduced earlier in this chapter. With these models, there is no intention to model any particular empirical case, and for some models, it may be difficult to find any close connection with observable data at all. For example, Schelling's model is usually built on a toroidal regular grid with agents dichotomized into two classes (e.g., red and green). These characteristics of the model are plainly not intended as empirical descriptions of any real city or real households. How then might such models be validated?

The answer is to see such models as part of the process of development of theory, and to apply to them the criteria normally applied to evaluating theory. That is, abstract models need to yield patterns at the macro level that are expected and interpretable; to be based on plausible micro-level agent behavioral rules; and, most important, to be capable of generating further, more specific or "middle range" theories (Merton, 1968). It is these middle range theories and the models based on them that may be capable of validation against empirical data. If an abstract model has been created using a deductive strategy, there will already be some hypotheses about the agent's behavior and about the macro-level patterns that are to be expected. The first validation test is therefore to assess whether the model does indeed generate the expected macro-level patterns. A more thorough test would be to see what happens when parameters of the model are systematically varied (see Section 3.4.1, on sensitivity analysis). One would hope that either the macro-level patterns persist unchanged with variation in parameters, or, if they do change, the changes can be interpreted. For example, in Schelling's model, one can alter the tolerance level of the agents. At low values of tolerance, households rarely find a spot where they are happy and the simulation takes a long time to reach a steady state, if it ever does. At sufficiently high values of tolerance, households are satisfied whatever the color of their neighbors and the initial random distribution hardly changes.

Once these basic tests have been passed, one can evaluate whether the model can be used to inform theories about specific social phenomena, and then those theories can be tested. For example, the Schelling model, although developed for ethnic segregation, is more general and abstract than this implies. It could be applied to any characteristic of actors. When used to explain ethnic segregation, however, one needs to start developing the theory to include the other factors that are of undoubted importance in location decisions in urban areas, including affordability and availability of the housing stock, the presence of more than two ethnic groups and people who belong to none or more than one, and the functional form of ethnic attitudes. (For instance, Bruch and Mare, 2006, suggest that the segregation effect in the Schelling model depends on the agents having a dichotomous attitude of being either happy or unhappy, and that clustering does not result if agents have a smoothly continuous attitude ranging from very unhappy to very happy.)

3.3.2 Middle Range Models

Models such as those mentioned in Chapter 1 that simulate consumer behavior, industrial districts, or innovation networks are intended as "middle range" simulations: They aim to describe the characteristics of a particular social phenomenon, but in a sufficiently general way that their conclusions can be applied widely to, for example, most industrial districts rather than just one.

The generic nature of such models means that it is not usually possible to compare their behavior exactly with any particular observable instance. Instead, one expects to be satisfied with qualitative resemblances. This means that the dynamics of the model should be similar to the observed dynamics and that the results of the simulation should reveal the same or similar "statistical signatures" as observed in the real world; that is, the distributions of outcomes should be similar in shape (Moss, 2002).

For example, the firms that one finds in innovation networks have collaborative links with other firms in the same industrial sector. If one counts the number of partners of each firm and plots the log of the number of partners against the log of the number of firms with that many partners, the graph is approximately a straight line with constant slope (e.g., Powell, White, Koput, & Owen-Smith, 2005, Figure 3). A linear relationship between logged variables is the statistical signature of a *power law,* and it is characteristic of many social networks, from utility power networks to the World Wide Web (Barabási, 2003). We would expect that a simulation of an innovation network would also show a power law distribution of inter-firm links with a similar slope.

An example of a middle range model is Malerba, Nelson, Orsenigo, and Winter's (2001) work on the computer industry. They describe their model as "history-friendly," by which they mean that while the model does not reproduce the exact history of the computer industry, it does

> capture in a stylized and simplified way the focal points of an appreciative theory about the determinants of the evolution of the computer industry. It is able to replicate the main events of the industry history with a parameter setting that is coherent with basic theoretical assumptions.

They also note that "changes in the standard set of parameters actually lead to different results, 'alternative histories' that are consistent with the fundamental causal factors of the observed stylized facts."

3.3.3 Facsimile Models

Facsimile models are intended to provide a reproduction of some specific target phenomenon as exactly as possible, often with the intention of using it to make a prediction of the target's future state, or to predict what will happen if some policy or regulation is changed. For example, a business may be interested in finding the consequences for their inventory level of reducing the interval between sending out restocking orders. It is likely to require a model that precisely represents all their suppliers, the goods each supplies, and the unit quantities of those goods in order to be able to make reasonable predictions. Another, very different example is the work by Dean et al. (1999) on the Anasazi Indians in the southwestern United States. These people began maize cultivation in the Long House Valley in about 1800 BC, but abandoned the area 3,000 years later. Dean et al.'s model aimed to *retrodict* the patterns of settlement in the valley and match this against the archaeological record, household by household.

If such exact matches can be obtained, they would be very useful, not only as a powerful confirmation of the theory on which the model is based, but also for making plausible predictions. However, there are reasons for believing that simulations that exactly match observations of specific phenomena are likely to be rare and confined to rather special circumstances. Most social simulations contain some element of randomness. For example, the agents may have initial characteristics that are assigned from a random distribution. If the agents interact, their interaction partners may be selected randomly, and so on. The same is presumably true of the social world: There is a degree of random chance in what happens. The effect of this is that running the model a number of times will yield different results each time (this is dealt with in more detail in the next section). Even if the results are only slightly different, the best one can hope for is that the most frequent outcome—the mode of the *distribution* of outputs from the model—corresponds to what is actually observed (Axelrod, 1997a; Moss, 2002). If it does not, one might wonder whether this is because the particular combination of random events that occurred in the real world is an outlier and, if it were possible to "rerun" the real world several times, the most common outcome would more closely resemble the outcome seen in the model!

3.4 Techniques for Validation

Two areas need to be examined when validating models: first, the fit between a theory and the model of that theory, and second, the fit between the model and the real-world phenomenon that the model is supposed to simulate.

3.4.1 Comparing Theory and the Model: Sensitivity Analysis

The fit between a theory and its model is best evaluated by using the theory to derive a number of propositions about the form of the relationships expected between variables and then checking whether the expected distributions do, in fact, appear when the model is run using a variety of parameter settings. Each of the parameter settings corresponds to an assumption made by the model. One should aim to check each of the settings either by measuring the value from empirical data or by conducting a *sensitivity analysis*. Although the former is preferable, there will be many parameters that cannot be checked empirically, and for these, some form of sensitivity analysis is essential.

Sensitivity analysis is aimed at understanding the conditions under which the model yields the expected results. For example, with the opinion dynamics model described in Chapter 1, one might ask, how extreme do the extremists have to be for all the agents eventually to join one of the extreme parties? To find out, one needs to run the simulation for a series of values of the uncertainty parameter, perhaps ranging from 0.5 to 1.0 in steps of 0.1 (i.e., six runs). But the model includes random elements (for example, the order in which agents "meet" and exchange opinions is random), so one should not be content with just six runs, but should perform a number of runs for each value of the uncertainty parameter to obtain a mean and variance. If one does 10 repetitions for each parameter setting (the number needs to be chosen with the amount of variation in mind, so that one gets a statistically meaningful result), we would need to carry out 60 runs.

To make matters worse, most models include many parameters, and their interaction may affect the simulation (e.g., the number of extremist groups that emerge depends on both the uncertainty of the extremists and the distribution of extremists across the political spectrum, the parameters having an effect both independently and in combination), so that ideally one would want to examine the output for all values of all parameters in all combinations. Even with only a few parameters, this can require an astronomical number of runs and thus is not a practical strategy.

If we think of the range of each parameter as lying on an axis, the set of all parameters defines a multidimensional parameter space, in which each point corresponds to one combination of parameter values. The scale of the task in doing a full sensitivity analysis can then be quantified as the volume of this space, and any way of cutting down the space will reduce the number of simulation runs needed. One obvious way is to use prior empirical knowledge to restrict the range of as many parameters as possible. For instance, we might know that a parameter, although theoretically capable of taking any value between 0 and 100, in fact is never observed to have a value greater than 10. Alternatively, we can limit the applicability of the model by constraining the range of values we test: We may state that the model applies only if the parameter is somewhere between 5 and 10 and not investigate what happens when it is outside this range.

Another approach, which can be used in combination with limiting the range of parameters, is to sample the parameter space. Instead of performing simulation runs at every point in the space, we use only some points. These points may be chosen randomly or purposively to inspect combinations that we think are particularly interesting or that are close to regions where major changes in the simulation's behavior are expected (*phase changes*).

A sophisticated version of this approach uses a learning algorithm, such as the genetic algorithm (see Section 5.2.2) to search the space to identify regions where some output variable or variables take their maximum or minimum values (Chattoe, Saam, & Möhring, 2000).

3.4.2 Comparing the Model and Empirical Data

As discussed in the previous section, not all models are expected to match empirical data; there may be no data available to compare with models whose objective is the development of theory, and no reason to conduct empirical tests. For middle range models, the criterion is whether the simulation generates outputs that are qualitatively similar to those observed in the social world, but a quantitative match is not expected. It is only with what we have called facsimile models that there are stringent requirements for comparison between data obtained from the simulation and empirical data. This section considers some of the ways of doing this.

Social scientists are well used to comparing data obtained from models and data collected from the social world: This is what is being done implicitly every time one calculates an R^2 (the coefficient of determination) in an ordinary linear regression (Fielding & Gilbert, 2000). The model in this case is the regression equation, which computes the predicted values of the dependent variable. Similarly, one can measure the fit between the values of a variable that are output from a simulation model and the values observed empirically (this is, in fact, just the correlation coefficient between the two sets of values). However, this simple procedure makes a number of strong assumptions that, although often satisfied for linear regressions, are much less likely to be appropriate for simulation models.

One important characteristic of simulation models is that the values of output variables change as the simulation runs. For example, in a model of consumer behavior, the number of purchasers of a particular brand might be observed growing from zero to a majority during a simulation. The growth trend might then be compared with the growth in sales of an actual product. Because the data to be compared are time series, one must allow for the fact that there is autocorrelation: The value at time $t + 1$ is not independent of the value at time t. Statistical procedures called ARIMA can be used to compare such time series (Chatfield, 2004).

3.5   Summary

In this chapter, a number of conceptual issues involved in designing and carrying out agent-based modeling research have been considered. We have shown by example the type of analysis that needs to be done before one begins programming and have mentioned some of the challenges that are raised by wanting one's model to be verifiable and valid. The next chapter will move on to matters of implementation: how one can code a model; plan its development; and, finally, report its results.

4. DESIGNING AND DEVELOPING
AGENT-BASED MODELS

4.1   Modeling Toolkits, Libraries, Languages, Frameworks, and Environments

Although some modelers build their agent-based models using only a conventional programming language (most frequently Java, although any language could be used), this is a hard way to start. Over the years, it has become clear that many models involve the same or similar building blocks with only small variations. Rather than continually reinventing the wheel, commonly used elements have been assembled into *libraries* or *frameworks* that can be linked into an agent-based program. The first of these to be widely used was Swarm (http://www.swarm.org/), and although this is now more or less completely superseded, its design has influenced more modern libraries, such as Repast (http://repast.sourceforge.net/) and Mason (http://cs.gmu.edu/ ~ eclab/projects/mason/). Both of the latter are written in Java and so link most easily to models that are also written in Java, but they can be used with other languages. Repast is available in a version for .NET and can be linked easily to programs in C#, Visual Basic, and Python. Both provide a similar range of features, such as the following:

- A variety of helpful example models
- A sophisticated scheduler for event-driven simulations
- A number of tools for visualizing on screen the models and the spaces in which the agents move
- Tools for collecting results to a file for later statistical analysis
- Ways to specify the parameters of the model and to change them while the model is running
- Support for network models (managing the links between agents)
- Links from the model to geographical information systems (GISs) so that the environment can be modeled on real landscapes.
- A range of debugged algorithms for evolutionary computation (Section 5.2.2), the generation of random numbers, and the implementation of neural networks.

Many person-years of effort have gone into the construction of these libraries, and using them can greatly reduce the time taken to develop a model and the chance of making errors. Both Repast and Mason are open-source software, available free for noncommercial use. Their only disadvantages are, first, their complexity, which means that it can take some months before one can take full advantage of the wide range of features they offer; and second, that the modeler is expected to use a relatively low-level language such as Java to develop his or her model.

More suited to the beginner are *modeling environments* that provide complete systems in which models can be created and executed, and the results visualized, without leaving the system. Such environments tend to be much easier to learn, and the time taken until one has a working model can be much shorter than it would be if one were using the library approach. However, the simplicity comes at the price of less flexibility and slower speed of execution. It is worth investing time to learn how to use a library-based framework if you need the greater power and flexibility they provide, but often simulation environments are all that is needed.

Environments primarily intended for other purposes can also be used for simulation, sometimes quite effectively. For example, simple simulations can be created using the spreadsheet package Microsoft Excel, and the free, open source statistics package R (http://www.r-project.org/) can be useful for models that involve processing large amounts of data. Several significant agent-based models have been constructed using the mathematical packages MatLab (e.g., Thorngate, 2000) and Mathematica (e.g., Gaylord & D'Andria, 1998). Nevertheless, an environment designed specifically for agent-based modeling is usually the first choice.

4.1.1 Repast

Repast (North, Collier, & Vos, 2006) is a family of two libraries, one for Java and one for Microsoft's .NET, and a more visual tool that allows developing simulations using the scripting language Python. The two libraries are functionally identical, and the choice of which one to use should depend only on your choice of programming language. The Python system is closer to a simulation environment and needs less advanced programming skills.

Repast has an active user base—from academia, government, and business—and a mailing list that is very helpful if you have questions about how to use it. The Java version will run on almost any platform, including Windows, Mac OS X, and Linux.

4.1.2 Mason

Mason (Luke, Cioffi-Revilla, Panait, Sullivan, & Balan, 2005) is another Java library, also influenced greatly by Swarm. It offers both standard and 3-D visualization libraries, and the user can record movies of the simulation as it runs (both of these features are also available in NetLogo; see below). Mason is free and open source.

4.1.3 NetLogo

Currently, the most popular agent-based simulation environment is NetLogo (Wilensky, 1999). It includes a user interface builder and other tools such as a system dynamics modeler. It is available for use free of charge for educational and research purposes and can be downloaded from http://ccl.northwestern.edu/netlogo/. It will run on all common operating systems: Windows, Mac OS X, and Linux. NetLogo has an active user community that answers users' questions quickly and thoroughly, and has users in all levels of education and in the natural as well as the social sciences. Other environments include StarLogo (http://education.mit.edu/starlogo/) and AgentSheets (http://agentsheets.com/), but these are more suited to creating and demonstrating very simple models for teaching than building simulations for research.

4.1.4   Comparison

Table 4.1 compares Swarm, Repast, Mason, and NetLogo on a number of criteria, using admittedly subjective judgments. None is ideal for all uses. To choose between them, one needs to consider one's own expertise and experience in programming, the likely complexity of the model, and the aims of the project (e.g., is the project very exploratory and the model likely to be fairly simple, or does the project intend to build a relatively complicated model and exhaustively test its behavior against data?). Repast, Mason, and NetLogo continue to be developed and are evolving quite rapidly, so the information in Table 4.1 needs to be checked against the current state of each of the systems. Other comparisons and reviews can be found in Castle and Crooks (2006), Gilbert and Bankes (2002), Railsback, Lytinen, and Jackson (2006), and Tobias and Hofmann (2004).

NetLogo stands out as the quickest to learn and the easiest to use, but may not be the most suitable for large and complex models. Mason, which is comparable to, but often faster than, Repast, has the advantage of being the newest, drawing on the experience of the older systems, but also has a significantly smaller user base, meaning that there is less of a community that can provide advice and support.

4.2   Using NetLogo to Build Models

In the remainder of this book, we shall use the agent-based simulation environment NetLogo (Wilensky, 1999). NetLogo, like the other environments and libraries mentioned above, is undergoing continuous development, with a major new version appearing more than annually. Hence, the code printed in this book (which was developed using version 3.1) may need modification to run in future versions. However, Wilensky and his team strive to make changes to NetLogo upwards compatible, so any changes needed may be made automatically when you load the code, or may require only minor editing.

The NetLogo system presents the user with three tabs: the Interface tab, the Information tab, and the Procedures tab. The Interface tab is used to visualize the output of the simulation and to control it (see Figure 4.1), the Information tab provides text-based documentation of what the simulation

TABLE 4.1
A Comparison of Swarm, Repast, Mason, and NetLogo

	Swarm	*Repast*	*Mason*	*NetLogo*
*License**	GPL	GPL	GPL	Free, but not open source
Documentation	Patchy	Limited	Improving, but limited	Good
User Base	Diminishing	Large	Increasing	Large
Modeling Language(s)	Objective-C, Java	Java, Python	Java	NetLogo
Speed of Execution	Moderate	Fast	Fastest	Moderate
Support for Graphical User Interface Development	Limited	Good	Good	Very easy to create using "point and click"
Built-in Ability to Create Movies and Animations	No	Yes	Yes	Yes
Support for Systematic Experimentation	Some	Yes	Yes	Yes
Ease of Learning and Programming	Poor	Moderate	Moderate	Good
Ease of Installation	Poor	Moderate	Moderate	Very easy
Link to Geographical Information System	No	Yes	Yes	No

*GPL: General Public Licence, http://www.gnu.org/copyleft/gpl.html

is for and what should be observed, and the Procedures tab is where one writes the simulation program using a special language specific to this environment (the NetLogo language). NetLogo is based on the programming language Logo (Papert, 1980), which was designed for teaching young children about the concepts of procedures and algorithms and was originally used to control small toy robots called "turtles." In memory of this, NetLogo's agents are still called turtles.

The Interface tab includes a black square called the *view*, which is made up of a grid of *patches*. This is the spatial environment in which the agents move: A simulation program can instruct agents to move in any direction from patch to patch, and the agents will be visible on the view (see, for

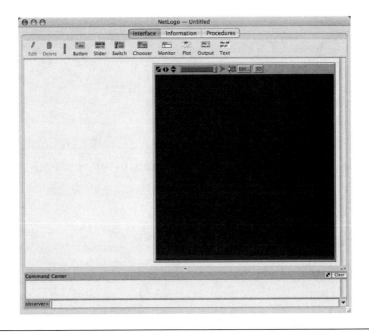

Figure 4.1   The NetLogo Interface

example, Figure 4.3; the small triangles on the view are agents). Usually, the NetLogo environment is configured so that the left-hand edge joins on to the right-hand one and the top edge to the bottom, so that if an agent moves off the left-hand side of the view, it immediately reappears at the right-hand side (the environment is topologically equivalent to the surface of a toroid, a donut-shaped solid). Patches start colored black but can easily be recolored, so that, for example, one could create a contour map. The number of patches in the view can also be configured: When NetLogo starts, the view consists of 35 by 35 patches, but the number can be increased to many thousands.

A NetLogo program has three parts. First, there is a section that says what kinds of agents there will be and names the variables that will be available to all agents (the *global variables*). Second, there is a setup procedure that initializes the simulation. Third, there is a go procedure, which is repeatedly executed by the system in order to run the simulation. Figure 4.2 shows a very simple example to give a flavor of a NetLogo program in which 10 agents are created and move randomly around indefinitely.

In this program, there are no global variables, so the program starts with the setup procedure. Any turtles left from a preceding run are cleared away, and

52

```
to setup
  clear-all
  create-turtles 10
end
to go
  ask turtles [
    right (random 360)
    forward 1
    ]
end
```

Figure 4.2   A NetLogo Program to Create 10 Agents and Make
Them Move at Random

10 new turtles are created (these are placed in the center of the NetLogo view).
The go procedure tells each turtle (agent) to carry out the commands within
square brackets: first, turn to the right (i.e., clockwise) by a random number of
degrees, and then move one unit forward, where the unit is the length of the
side of a patch. Each turtle moves independently of the others, all at the same
time (because NetLogo runs on an ordinary computer, the agents cannot all
operate at precisely the same time, but NetLogo makes it look as though they
do by using an asynchronous random update; see Section 2.3).

To run this program on your own computer, you would need to download
and start up NetLogo. Then click on the Procedures tab and type in the lines of
code shown in Figure 4.2. Move back to the Interface tab. Click on the Button
icon at the top and then on the white area next to the view. NetLogo will draw
a button that you should label "Setup." Then do the same for a Go button, also
setting the check box Forever on (this will cause the go procedure to be executed in a loop when the Go button is clicked, continuing until the button is
clicked a second time to stop the run). Clicking on your Setup button will create 10 turtles, shown piled on top of each other in the center of the view. Clicking on the Go button will send the agent turtles darting around the view,
following a random trajectory. Click on the Go button again to stop the program. The NetLogo interface should then look similar to Figure 4.3.

Although this is a very simple example, it does give an idea of how
quickly one can develop agent-based simulations in an environment like
NetLogo. The graphics for buttons (and sliders, switches, etc.) to control a
simulation are available through "drag and drop." The view offers many
possibilities for the visualization of agents and their environment without
doing any programming. NetLogo will also show dynamically changing
plots of output variables on the Interface tab. Although the NetLogo
programming language is somewhat different from the usual procedural

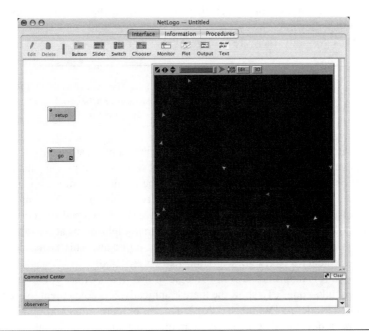

Figure 4.3   The Simple Program Running

languages, it is both powerful and (mostly) elegant, with the result that complex simulations can be programmed in surprisingly few lines of code.

There is no space in this book to provide a detailed tutorial on NetLogo, but the system includes a good built-in tutorial (find it under the Help menu) and comes with a large number of demonstration and example models, some of which are relevant to social science.

4.3   Building the Collectivities Model Step by Step

In Chapter 3, the "collectivities" model was introduced. In brief, this is a model that simulates the dynamic creation and maintenance of knowledge-based formations such as communities of scientists, fashion movements, and subcultures. The model's environment is a spatial one, representing not geographical space, but a "knowledge space" in which each point is a different collection of knowledge elements. Agents moving through this space represent people's differing and changing knowledge and beliefs. The agents have only very simple behaviors: If they are "lonely," that is, far from a local concentration of agents, they move toward the crowd; if they are crowded, they move away.

Thus, formally, there are two agent behavior rules:

Condition	Action
The agent is lonely	Move toward the crowd
The agent is crowded	Move away from the crowd

The first step in building the model is to make some basic decisions about the agents and the environment. The model specification implies that there will be only one type of agent and that the agents will move about in a space. We need to decide the dimensionality of this space: For simplicity, we shall use a two-dimensional grid that can be mapped directly onto NetLogo's view. To avoid special effects that might occur at the edges of the grid, we shall use a toroid, which has no edges. This is the default arrangement for NetLogo, so nothing extra is required.

Next, it is helpful to lay out the logic of the model, either graphically or in "pseudo-code." To show the logic graphically, it is convenient to use the Unified Modeling Language (UML), a means of representing programs that has been developed as a way of communicating software independently of the details of programming languages (Miles, 2006). UML provides a range of standardized diagrams that can be used to show the class hierarchy of the objects in the program; a sequence diagram that shows how one thing leads to the next; and an activity diagram, which is similar to a flowchart. UML is very good for describing a model in, for example, a published paper, but the collectivities model is so simple that UML is hardly necessary. For instance, there is only one class (for the agents) and only two agent actions (move forward and turn around).

With such a simple model, an alternative approach is more helpful: to use "pseudo-code," an informal mixture of natural language and programming conventions that makes the structure and flow of a program clear without requiring the reader to be familiar with any particular programming language. Figure 4.4 shows the collectivities program in pseudo-code.

The program is in two parts: the initialization (called "Setup" in NetLogo) and the execution ("Go" in NetLogo). The indentation of the pseudo-code helps to clarify which lines go with which. For example, the program loops repeatedly carrying out the lines between Loop forever and End loop. Constant parameters of the model are shown in *italics*.

Once one has a pseudo-code version of the program, it is relatively easy to translate it into a programming language such as NetLogo, and the results of doing so are shown in Figure 4.5. Figure 4.6 is a screen shot of the

Initialization

```
Create agents and distribute them at random in
  knowledge space
```

Execution

```
Loop forever
  Each agent:
    Counts the number of other agents within its
      local-radius
  Each agent:
    Compares the number of other agents within its
      local-radius with the threshold
    If the number is greater than the threshold
    Then (the agent is crowded)
      The agent locates that agent within
        visible-radius with the most agents
        surrounding it
      The agent moves a distance proportional to
        speed away from this agent
    Else (the agent is lonely)
      The agent locates the agent within
        visible-radius with the most agents
        surrounding it
      The agent moves a distance proportional to
        speed towards this agent
End loop
```

Figure 4.4   The Collectivities Program in Pseudo-Code

program code in the NetLogo environment, and Figure 4.7 shows the user interface with buttons to set up the simulation (pressing this button executes the initialization code, from setup to the end statement 10 lines further on) and to run the simulation (Go, which executes the go procedure repeatedly until the button is pressed again).

4.3.1   Commentary on the Program

In this section, each line of the program in Figure 4.5 is explained.

```
breed [agents agent]
```

This line names the class that will be used in the model, in both its plural and singular forms.

```
agents-own [ around visible ]
```

```
breed[agents agent ]
agents-own[ around visible ]
to setup
  clear-all
  ask patches[set pcolor white ]
  create-custom-agents 500 [
    set color green
    set size 2
    ; distribute agents randomly
    setxy random-pxcor random-pycor
    set heading random 360
    ; ensure that each is on its own patch
    while[any? other-agents-here] [ fd 1 ]
    ]
end

to go
  ask agents [ count-those-around ]
  ask agents [ move ]
end

; store the number of agents surrounding me within
; local-radius units
; and the agents that I can see within visible-radius

to count-those-around
  set around count agents with[self != myself] in-radius
    local-radius
  set visible agents with[self != myself] in-radius
    visible-radius
end

to move
  if any? visible [
    ifelse around > threshold
      [ ; if more than threshold agents surround me,
        I'm crowded:
        ; face away from the most popular patch and
          become red
        face-away set color red ]
      [ ; else I'm lonely: face towards the most popular
        spot and
        become green
        face-towards
        set color green ]
    ]
  ; and move in my (new) direction
```

```
    fd random speed
    ; ensuring that I stop on an empty patch
         while [any? other-agents-here] [ fd 1 ]
end

; face towards the most popular local spot
to face-towards
    face max-one-of visible [around]
end
; face away from the most popular local spot
to face-away
    set heading towards max-one-of visible [around] - 180
end
```

Figure 4.5   The Collectivities Program

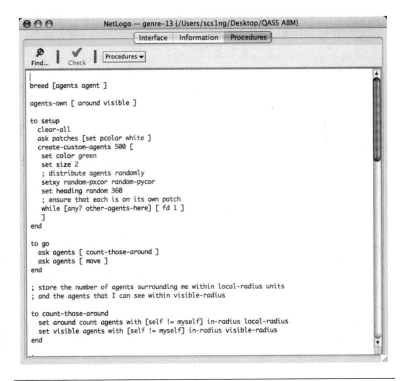

Figure 4.6   The NetLogo Procedures Tab Showing Part of the
Collectivities Program

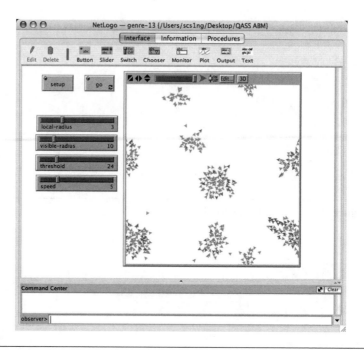

Figure 4.7   The Collectivities Interface

The agent class has two attributes, `around` and `visible`. Every agent has its own values for these variables, which store the number of other agents surrounding the agent and the set of agents that it can see, respectively.

```
to setup
  clear-all
  ask patches [set pcolor white ]
```

The setup procedure is executed when the user presses the Setup button on the interface. The procedure first deletes any agents left over from a previous run and then colors all the patches on the grid white. The code `ask patches [. . . ]` tells all the patches (recall that a patch is a cell on the grid) to execute the code inside the square brackets. The term `pcolor` is a NetLogo variable that sets the color of a patch.

```
create-custom-agents 500 [
  set color green
  set size 2
```

Then 500 agents are created. The commands inside the square brackets, are executed by each of the new agents independently. The image of each agent on the view (see Figure 4.7) is colored green, and its size is set to twice the minimum.

```
setxy random-pxcor random-pycor
set heading random 360
```

Then each agent is moved to a random spot on the grid (random-pxcor yields a number corresponding to a random position between the left and right edges of the grid, and random-pycor does the same for a position between the top and bottom edges), and the heading (the direction that the agent faces) is also set to a random direction.

```
while [any? other-agents-here] [ fd 1 ]
```

It is possible that the result of the random distribution of agents will be that two or more agents end up on the same patch, one standing on the head of another, so this line gets each agent to check whether there are any other agents in the same patch. If there are, the agent advances by one unit (fd 1) in the direction of its current heading and checks again, continuing to advance until it has found an unoccupied patch.

```
    ]
end
```

This completes the commands that are executed by each agent immediately after it is created, and also brings the setup procedure to an end.

```
to go
    ask agents [ count-those-around ]
    ask [ move ]
end
```

The go procedure gets executed repeatedly when the Go button on the interface is pressed. Each time through the procedure, the simulation moves on one time step. At each step, all the agents are asked to count the number of agents that are in their local area on the grid, and then all the agents are asked to move. It is important that all agents complete counting the agents around them before any agents move; otherwise, the counts may be confused by agents moving before they have been counted. NetLogo guarantees that all agents complete the commands in a construction such as ask agents [. . .] before the next command (in this case, the second ask agents [. . .]) is started. This would not have been the case if the program had been written ask agents [count-those-around move] because then each agent would have started to move as soon as it had finished counting, and for some agents, this may have occurred before other agents had added them to their counts. This is an example of the kind of timing problem that one needs to guard against when programming agents that appear to act simultaneously.

The `count-those-around` and `move` procedures are both methods of the `agent` class, that is, procedures that are defined in the program code to show what it means for agents to count and move. The next lines specify what `count-those-around` consists of.

```
to count-those-around
    set around count agents with [self != myself]
      in-radius local-radius
    set visible agents with [self != myself] in-radius
      visible-radius
end
```

In the first line of this procedure, the variable `around`, owned by each agent, is set to the number of agents that are located within a radius of `local-radius`. The user sets the value of `local-radius` before the simulation starts by moving a slider on the interface (see Figure 4.7). It is easiest to see what this line of code does by working backwards from the end. The code `in-radius local-radius` yields those agents that are located at or within a distance of `local-radius` away from the agent executing this procedure. This set of agents includes the agent itself (it is zero units away from itself, and therefore closer than `local-radius`). The code `with [self!=myself]` excludes that agents, and `count` returns the number of agents remaining. The agent's attribute `around` is set to this number. One can think of the agent looking around, counting the number of agents it sees within the `local-radius` and remembering this in the `around` variable. In the terminology of production systems (see Section 2.1.2), `around` is part of the working memory.

The next line of code is very similar. It finds which other agents can be seen by the agent, that is, which agents are within the `visible-radius`, a parameter whose value is set by another slider on the interface. The second line is slightly different from the first because the variable, `visible`, stores not a count of the visible agents but the names of the agents themselves. This is necessary so that they can later be identified (in procedures `face-towards` and `face-away`).

The `move` procedure both decides the direction in which the agent is to move and performs the action. In terms of production systems, the `move` procedure is a very simple rule interpreter.

```
to move
    if any? visible [
```

The procedure begins by checking whether there are any agents visible to the agent executing this method. If there are, the agent needs to decide whether it is "lonely" or "crowded," that is, it needs to decide whether either of the condition-action rules apply.

```
ifelse around > threshold
    [
        face-away
        set color red ]
```

The condition consists of checking whether the number of agents around this agent is greater than the threshold parameter set with a slider on the user interface. If it is, it is "crowded" and the agent must turn away from the center of the crowd. The procedure face-away does this. In addition, to show these agents on the view, their color is changed to red.

```
    [
        face-towards
        set color green ]
    ]
```

The alternative is that the agents are "lonely," and they must turn toward the crowd and become green.

```
fd random speed
while [any? other-agents-here] [ fd 1 ]
end
```

Once the agent has set its direction, it can move, at a rate determined by the value of the speed parameter (random speed returns a random number less than the value of speed). As before, if the agent lands on a patch that is already occupied by another agent, it continues to move forward in the same direction until it finds an empty patch.

That concludes the move procedure. We still have to define face-towards and face-away.

```
to face-towards
    face max-one-of visible [around]
end
```

The face-towards command retrieves the set of agents that this agent can see (the agents that have been remembered in the visible variable) and finds how many agents surround each of those agents. One of the agents with the most agents around it (max-one-of visible [around]) is considered to be at the center of the crowd toward which this agent wants to move. The NetLogo command face sets the heading of the agent to the direction of that central agent.

62

```
to face-away
  set heading towards max-one-of visible [around]  - 180
end
```

The `to face-away` command is almost the same, but the direction is 180 degrees opposite to the direction of the most central agent.

That concludes the program code for the collectivities model.

Running the model shows that the initial uniform random distribution of agents separates into "clumps," in which some agents are central and others are distributed around them. The central agents are crowded, and so move. In doing so, they shift the centroid of the clump slightly and may make other agents either crowded or lonely, and they too will move. Thus, the clump of agents, although remaining together for long durations (as measured in time steps), drifts across the view. Lonely agents move toward the clump, sometimes joining it and sometimes continuing to trail behind it. The clumps never merge.

Figure 4.8 illustrates a typical snapshot. In this figure, agents that are crowded are a dark grey and those that are lonely are a lighter grey.

Comparing the behavior of this model with the features of collectivities described in Section 3.1, we see the following:

1. When we run the model, we see "clumps," but drawing a boundary around clumps involves some arbitrary definition, perhaps in terms of local densities of agents.

2. Although a definition of which agents are in and which are out of a clump is possible (e.g., in terms of the distance to the nearest neighbor), again this seems arbitrary.

3. Agents in the same clump are close together and so could be thought of as sharing some aspects of their knowledge.

4. The location of the clump, as indicated by the position of its centroid, is constantly changing as some agents move more closely into the clump and others seek new, less crowded locations.

5. Some agents consider themselves to be crowded, and these behave differently from the other agents in the clump (by "innovating" or trying to find less crowded positions by moving through the knowledge space). These agents are located more centrally in the clumps and are influential in setting the direction of travel of the other agents.

The features of collectivities that we observe in society thus emerge in the model as a result of the behavior of the agents. Although other micro-level

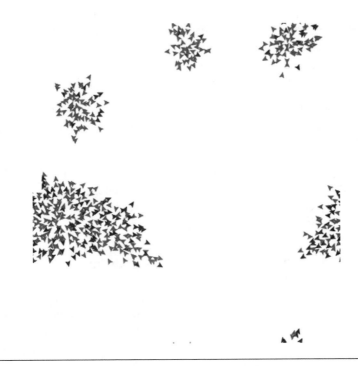

Figure 4.8   Snapshot of the Simulation

actions could produce the same or similar macro-level patterns (Gilbert, 2002), it is useful to know that these do yield the macro-level behavior that we observe. Specifically, we can conclude from the model that if

1. agents change their ideas in knowledge-space in response to "over-crowding" (Mulkay & Turner, 1971),

2. some ideas and some agents are considered to be high status or important, and

3. agents are motivated to copy and adopt those ideas or a variation on them,

then the phenomenon we have described as a "collectivity" will emerge from the agents' behavior.

A more detailed comparison of the model output with empirical data is not appropriate for this abstract model (see Section 3.3.1). The value of abstract models is twofold: They can account for the generation of particular phenomena, following Epstein's maxim that "to explain macro-scopic social patterns, we generate—or 'grow'—them in agent models" (Epstein, 2007, p. 50; see also Epstein, 1999); and they can help to highlight commonalities and differences between phenomena that otherwise might be considered incomparable. A better criterion for this model's success is therefore the degree to which it generates further theoretical questions or informs middle range theory that can be empirically validated. For exam-ple, the model suggests the question, What are the significant similarities and differences between "punks" and "scientists," given that their social formations can be re-created using the same generic model? It does seem that there are many areas of social life where similar micro-behaviors may be found and correspondingly many emergent collectivities, and so this model may be applied to account for a wide range of social phenomena.

4.4   Planning an Agent-Based Modeling Project

As with any research project, it is helpful to plan a simulation project step by step in advance. Then you can be more confident that what you plan is likely to be achievable, and remedial action can be taken if it becomes clear that you are falling behind schedule. Although most simulation projects are not different in their essentials from projects using other styles of research, there are some special features that need attention.

• *Do Not Underestimate How Long It Takes.* It is tempting to think only about the time spent writing the code, but it often takes as long to design a model as to code it, and frequently longer to debug a program than to write it. Therefore, it is not being too pessimistic if one estimates the time needed to write a program and then multiplies this by at least three to give the total time for model development. Unless one is using a modeling environment such as NetLogo, most of the programming will be taken up with the devel-opment of the user interface and output display routines, not with coding the model itself. This is one reason why modeling environments are so valuable—they can save a great deal of work.

• *Keep a Diary.* Ideas will occur to you at all stages of the project, and you risk forgetting them unless you jot them down in a diary or lab book. Pay special attention to problems that you encountered while building the model: Difficulties that you initially assume are just technical programming problems may turn out to have a wider significance. For example, if it seems that the results of the simulation are very sensitive to the particular value of a parameter, this may just be an issue in building the model, but it may also suggest some substantive conclusions about the role of this parameter in the real world.

There are some additional points that need consideration in larger projects, where there are several researchers working in a team.

- *Find People With Appropriate Skills.* If you are a lone researcher, you will know the extent to which you are already skilled in programming models. If the project is a larger one in which there is some division of labor, you will probably need to recruit people with expertise in modeling. Because agent-based simulation is still a new approach, researchers with significant experience are hard to find, and you may need to be content with hiring people with other skills and training them in agent-based modeling. Particularly useful skills are a familiarity with programming in Java (even if you are not intending to use one of the Java-based libraries—see Section 4.1—the grounding in programming that a Java course gives is very useful); some experience in researching in the domain to be modeled; and the ability to write clearly, which is vital for preparing reports and papers.

- *Attend to Intraproject Communication.* If more than one person is working on a project, attention needs to be paid to making sure that everyone understands each other and is aware of what the others are doing. Although this is true in all team tasks, in modeling projects, it is usual to have some people who are domain experts but know little about modeling, and others who are modeling experts but know relatively little about the domain. Both sides may feel inhibited about asking questions and exposing their ignorance to the others. In larger projects, it may be worth scheduling specific training sessions, where those with greater knowledge of particular aspects of the project teach the others in order to bring everyone up to a common level of knowledge and skill.

- *Pay Attention to Scheduling.* Most modeling projects involve some data collection and some model development. This can be tricky to schedule if the specification of the model awaits the collection and analysis of empirical data, but the collection of data rests on the prior definition of precisely what is to be measured. Unless care is taken, one can get into a Catch-22 situation, in which neither modelers nor data collectors can make a move.

4.5   Reporting Agent-Based Model Research

The best way of learning how to report the results of agent-based modeling is to study how others have done it. Take a sample of papers that you have found helpful or interesting and look closely at how the authors constructed them and what makes them persuasive. Although agent-based modeling is too young an area to have a well-developed set of conventions about how papers should be written, there are some common elements (Axelrod, 1997a). A helpful discussion can be found in Richiardi, Leombruni, Saam, and Sonnessa (2006).

The main sections of an agent-based modeling report or journal article are usually as follows:

1. An abstract. This should indicate (in roughly this order)

   a. The main research question considered in the paper
   b. The findings and conclusions of the paper
   c. The methods used (e.g., agent-based modeling; survey analysis) and, for empirical data, the sample from which data were collected

2. An introduction that sets out the background to the issue addressed in the paper and explains why it is of interest.

3. A literature review that discusses previous work and shows why the research reported in this paper is a worthwhile addition or improvement to the prior work. Both literature on the research problem or domain, and on related models, even if these have not previously been applied to the domain, should be reviewed. This section should make clear in which respects the reported research is an advance and how it is using previous work.

4. A statement of the regularities that you want to explain (this will usually be a summary of the material in the introduction and review). These may be stated as a set of formal hypotheses that you aim to (dis)prove, or they may be presented less formally.

5. A description of the model. The description needs to be sufficiently detailed that a reader could, in principle, reimplement your model and obtain the same results, but it should not include program code (some readers will not know or understand the programming language you have used). Instead, use diagrams (e.g., UML) or pseudo-code to describe your model (see Section 4.3). Pay particular attention to the sequence in which events occur in the model: This is the most frequent source of problems in reimplementing a model accurately. Do not be afraid of including equations relating variables if these will help to specify your model precisely.

6. A description of the parameters. The values you have chosen for each of the parameters need to be explained and justified. Some may be

based on observations of the social world (e.g., the employment rate in a model of the labor market); some may be plausible guesses, and you have investigated the effect of varying their values using a sensitivity analysis; and some may have been inferred "backwards," because it is only these values that give the patterns of output that you want to demonstrate with the simulation. All this needs to be explained.

7. A description of the results. This will almost certainly involve presenting and commenting on graphs that show how variables that you have observed from the simulation runs are related. Be careful to be clear about the conditions under which these simulations were carried out. For example, do the plots show averages of several runs, and if so, how many runs and how much variation was there between runs (you might consider using error bars to show the degree of variation)? If you are showing the trend in a variable over time from step zero onward, make sure that you have plotted a run long enough that it is clear that the trend has become established and is unlikely to change drastically just off the graph (Galan & Izquierdo, 2005). If you are relating the values of variables as they are at a particular time step, make sure that you state the time step at which the measurements were made.

8. A discussion of what steps you took to verify (see Section 3.2) and validate (see Section 3.3) the model, and what confidence the reader should therefore place in your results.

9. A conclusion. This should take the hypotheses listed in (4) and clearly state whether the model suggests that they are true, false, or not proven. This section can then develop the ideas from the introduction, proposing a general conclusion and perhaps speculating about the implications (e.g., if the paper is about the labor market, what policies might or might not be successful in reducing rates of unemployment).

10. Acknowledgments. Brief thanks to sponsors, funders, and those who have helped you to do the research.

11. A list of references containing only those works cited in the paper and no others. As always, you need to be sure to provide full bibliographic details in the format required by the journal in which you hope to publish.

12. Optionally, an appendix in which large tables and possibly the pseudo-code version of your model is placed.

4.6   Summary

This chapter has described the process of implementing an agent-based model, from choosing a toolkit to reporting on the results. The most important step is to plan your work before you get too engaged with it. This will be helpful in ensuring that you have a well-defined research question to answer, and that you have allowed sufficient time and resources to be able to answer it.

5. ADVANCES IN AGENT-BASED MODELING

Agent-based modeling is a rapidly advancing field, and new approaches are being introduced all the time. In this chapter, we shall briefly describe some of the directions of current research.

5.1   Geographical Information Systems

Most agent-based models involve agents moving over a landscape, but usually this terrain is a rectangular plane or a toroid. In place of these abstract surfaces, newer simulation environments are offering the possibility of creating complex artificial surfaces, or incorporating terrains mapped from real landscapes. They do this by integrating a *geographical information system* (GIS) into the model. A GIS is a software system for managing, storing, and displaying spatial data such as maps (Chang, 2004; Heywood, Cornelius, & Carver, 2002). GIS study is a research specialty in its own right, and this is one of the reasons why using a GIS for agent-based modeling has been slow to develop: GISs have their own software technologies and techniques that have been somewhat difficult to merge with the tools of agent-based modeling.

GISs store spatial data in specially built or adapted "spatially aware" database systems. These are designed to be efficient at answering the kind of queries that one needs for managing geographical data, such as "tell me about all the other objects that are no more than 10 units away from this object." GIS data are often arranged in *layers*, containing data on one or a few variables. When displaying or manipulating a map, one can turn some layers on or off, to see just the variables in the visible layers. For example, a map might include roads and lakes in separate layers. If one wanted to see the road network, but was not interested in the lakes, the lake layer could be turned off. There are two kinds of GIS variables: raster and vector. Rasters are regular grids of cells, each cell storing a single value. Raster variables are used for data that vary continuously over the surface, such as height and hours of sunshine. Vector variables are used for data that are in the form of points, lines, or areas (known in GISs as *polygons*). Spatial data may be displayed using a projection, which is a way of showing the surface of a three-dimensional body such as the earth on a two-dimensional map. There are a wide variety of projections available, although the question of which projection to use is more likely to trouble a geographer wanting to map the whole earth than a modeler who wants to map just a town or region, small enough that the differences between alternative projections become irrelevant.

With a GIS, it is possible to build an agent-based model in which the agents traverse a more realistic landscape, for example, moving only along city streets. This is vital for projects such as modeling traffic flows (Raney, Cetin, Vollmy, & Nagel, 2002) and the spread of epidemics (Dibble & Feldman, 2004; Dunham, 2005), although it may be a distraction if the intention is to model an abstract process not specifically located in any particular place. One decision that needs to be made early is whether it is sufficient for the GIS landscape to be unchanging throughout the simulation run, or whether the landscape needs to change dynamically. For example, a static representation of a city is likely to be sufficient for a traffic model because in the time period represented by the simulation (a few hours or days), the arrangement of city streets will not change materially. On the other hand, a model of the effects of a hurricane on a city population may need to have its topography updated during the simulation to account for floods, closed streets, and storm damage. However, managing time-varying data is at the state of the art for GISs and a challenge if one wants to integrate it with agent-based modeling.

A second issue that needs to be considered when designing spatially aware agents is how they will detect the features of the terrain they are traversing. For example, if one wants to ensure that pedestrians walk down streets, and not through buildings, there needs to be a way for agents to determine that the way ahead is a walkway. This is best done by sending a query to the GIS to determine the locations of the objects ahead of the agent and then decomposing ("parsing") the returned answer to check that forward movement is not obstructed. In principle, this is not complicated, but in practice, it can take a lot of processing by both the underlying GIS and the agents and may slow down the model. These are problems with which the current generation of model builders is struggling, but over the next few years, they will be solved, and spatially aware agent-based modeling frameworks will become more common and easier to use.

5.2   Learning

The agents in most of the models mentioned so far are not capable of learning from experience. Simple agents driven by production rule systems have a memory in which their current and past state is recorded, so that they "learn" about the state of the environment as they proceed through it. But usually we mean something much more than this by learning. One kind of learning is learning more effective rules, and then changing behavior as a result of the learning. In an earlier section, we touched briefly on one technique that allows this kind of learning; although neural networks (see Section 2.1.3) are not based on explicit rules, the learning that they perform can be akin to rule learning. There are two other common techniques for modeling learning in agents: reinforcement learning and evolutionary computation. Both have been transferred from artificial intelligence, where the development of learning methods is an important topic under the heading of *machine learning*.

5.2.1   Reinforcement Learning

Reinforcement learning (RL) is akin to what we do when exploring a new city to find a tourist attraction by trial and error: You try one street, and if it looks promising you carry on, or if not, you double back to try another. Agents that engage in RL have a state (e.g., their current location) and a small set of possible actions (e.g., move one cell north, east, south, or west). The environment is set up so that it will eventually provide a reward (gratification in reaching the attraction). The agent has to find a *policy* that maximizes the reward it receives (Sutton & Barto, 1998). Usually, each additional step along the way to the goal reduces the eventual reward and so the goal is not just to find the reward, but also to do so in the most efficient way. The agent can try many times to find the best policy, exploring different routes, and so one of the issues in designing RL algorithms is the balance between exploration (trying new routes, which may be less efficient and waste time) and exploitation (keeping to the old routes, which may not be the best way). RL has been used to model firms that learn about the preferences of their customers (Sallans, Pfister, Karatzoglou, & Dorffner, 2003) and decision making in organizations (Sun & Naveh, 2004), among other phenomena. Several authors have used simple models of reinforcement learning to explain and predict behavior in the context of experimental game theory with human subjects (Chen & Tang, 1998; Erev & Roth, 1998; Flache & Macy, 2002; Macy & Flache, 2002; Mookherjee & Sopher, 1994, 1997).

RL is well suited to problems in which the environment and the reward structure remain constant during the simulation run. On the other hand, it does badly if the environment is dynamic (in particular, if all the agents are learning and the actions of one agent affect the state or the rewards of other agents, RL is not likely to be a good technique to use).

5.2.2 Evolutionary Computation

A very different approach to learning is that of evolutionary computation, a family of techniques of which the simplest and best known is the *genetic algorithm* (and others include evolutionary programming, evolution strategy, *genetic programming,* and *learning classifier systems*). Evolutionary computation (Eiben & Smith, 2003; Engelbrecht, 2002) is loosely based on natural selection and involves two basic processes: selection and reproduction. Fundamental to evolutionary computation is a population whose members reproduce to form successive generations of individuals who inherit the characteristics of their parents. The probability of successful reproduction is determined by an individual's *fitness*: fit individuals are more likely to breed and pass on their characteristics to their offspring than unfit ones.

The individuals involved in a genetic algorithm (Holland, 1975) may, but need not, be the agents in the simulation. For example, in a model of an industrial sector, the agents could be firms, with the sector as a whole learning how to be productive in the industrial landscape through successive generations of firm bankruptcies and start-ups (Marengo, 1992). On the other hand, each agent may learn using a genetic algorithm that acts on the agent's rules, each rule being an "individual" as far as the genetic algorithm is concerned (this is the way that learning classifier systems work) (Bull, 2004). Simulations of stock market trading have been built with such agents (Johnson, 2002; LeBaron, Arthur, & Palmer, 1999).

For a genetic algorithm, each individual must have its features encoded in a *chromosome*. The encoding may be in the form of a binary string, zero representing the absence of a feature, and one, its presence, or as a more complicated structure. It is also important that it is possible to assess the *fitness* of every individual—for example, the fitness may be measured by the accumulated capital stock of a firm agent, or the effectiveness of a cooperation strategy in a model of altruism. Individuals are selected for reproduction by the algorithm in proportion to their fitness, so that very fit individuals are very likely to breed and very unfit ones very unlikely. The reproduction takes place by combining the parents' chromosomes using a process called *crossover*: Slices are taken from each chromosome and combined to form a new one made of some sections from one parent's and some sections from the other's. In addition, to ensure that there continues to be some variation in the population even after a great deal of interbreeding, some bits from the offspring's chromosomes are randomly changed, or mutated.

The offspring's fitness is evaluated to determine its likelihood to reproduce to yield grandchildren. Eventually, the individuals who are relatively unfit die out and are replaced by individuals who have inherited from fit parents and are likely to be fit themselves. Although no individual does any learning as a result of the genetic algorithm, the population as a whole can be considered to be learning or optimizing as the individuals within it become fitter.

The genetic algorithm is usually a very effective optimization device. Whether it is a good model of any social phenomenon is more debatable (Chattoe, 1998). One difficulty is that it is often hard to see what would be an appropriate way of measuring "fitness," or even that the concept has any clear meaning when applied to social phenomena. Sometimes, this does not matter. For example, in a model of a simple society, one does not need a carefully elaborated definition of fitness: Agents can be designed to breed if they have the opportunity and have not previously died from lack of resources. Another common problem with the basic genetic algorithm is the difficulty of coding all the salient features of an agent into a binary string for the chromosome. However, other evolutionary computation techniques allow a wider range of structures. *Genetic programming* (Banzhaf, 1998; Koza, 1992, 1994), for example, substitutes program code for the binary string: Individuals evolve a program that can be used to guide their actions, and *learning classifier systems* (Bull, 2004; Meyer & Hufschlag, 2006) use condition-action rules as the equivalent of chromosomes.

5.3 Simulating Language

One of the defining features of multiagent models is that the agents have the potential to interact. It is this that separates multiagent models from, for example, microsimulation or equation-based modeling (see Section 1.4). The interaction may represent a simple perception of the presence of other agents, to avoid them or to imitate them, or it may involve more sophisticated communication of knowledge, opinions, or beliefs, depending on the requirements of the domain being simulated. However, even in the most sophisticated models, it is almost always the case that agent interaction occurs through unmediated and direct agent-to-agent message transfer. This is quite different from human communication, in which our thoughts (and intentions) have to be conveyed through an external natural language that is inevitably ambiguous and whose meaning has to be learned, rather than being innate.

As well as being unrealistic, the modeling of information transmission directly from agent to agent can give a misleading impression of a wide range of social phenomena, from cooperation to leadership, all of which change their character when modeled using error-free and unambiguous messages that go directly from one agent rule system to another. Partly to respond to this issue, and partly because how language evolved is an important topic in linguistics, research on the "evolution of language" has been increasing in volume and sophistication over the past decade (Perfors, 2002). Much of this work is specialized and complex; here we give only a rather simplified introduction to the topic and some hints about how language can evolve.

Suppose we have two agents, Adam and Betty, and that they are standing near to each other in a virtual landscape. Both can see a small red ball and a large blue cube, as well as a number of other objects nearby. Adam wants to direct Betty's attention to the red ball by making an utterance (implemented by transmitting a string of characters from one agent to the other). Initially, the agents do not share a language with which to speak about what they can see. Although Adam could utter some sequence of syllables, Betty would have no idea what this sequence meant and no way of knowing that Adam was mentioning the ball rather than the cube or any of the other objects they could both see. The problem is even more profound than this, because initially Adam has no concept of "ball" or "cube"; all that the agents start with is the ability to detect features such as color, shape, size, and location. Everything else has to be learned. The goal is to evolve a shared lexicon that is, in the words of Hutchins and Hazlehurst (1995), a "consensus on a set of distinctions" (p. 151).

The following is one algorithm that Adam and Betty could use (based on the work of Vogt, 2005, and Vogt & Coumans, 2003). Adam begins by looking at the ball and sees that it is round, red, and small. These features all distinguish the ball from the cube, which is square, blue, and large. Adam is able to categorize it as a "round, red, small" object, and Adam can henceforth discriminate such balls from square, blue, large objects. If there were two round, red, small objects in the scene, Adam would need to find another feature that discriminates the ball from the other object, for example, the objects' distance away from Adam. For Adam, "red" might be a range of shades: It would not be helpful to have a different category and thus a different name for every shade of red, so the color space is crudely cut into segments by the agent's perceptual system, and the same applies to the perception of other features such as shape and distance.

In this way, Adam is able to distinguish objects in his view, assigning each a bundle of categories chosen to differentiate the object from others previously encountered or currently present in the scene. The bundle of categories is called a *concept*. All of this takes place within the "head" of the agent. Betty will also have viewed the scene and will also have derived a concept for the ball, one that is likely to be similar to Adam's (but because they are standing in different places and so have different perspectives on the ball, Betty's concept may not be identical to Adam's). Adam now has to signal "ball" to Betty using an utterance. Assuming that this is the first time that Adam has spoken about balls, he invents a word for his concept—any string of characters will do, provided that it differs from any other word he knows. Adam "utters" the word and it is communicated to Betty. However, Betty has no idea what the word means at this stage.

In order to learn a common vocabulary, Adam and Betty engage in a *language game* (Steels, 2004). This involves repeated dialogues in which a score measuring the associations between all of Betty's concepts and the word in question are updated according to whether she judges that she has successfully understood the utterance. This judgment can be based on some nonverbal reinforcement (e.g., Adam pointing at the object, Betty succeeding in an action that depends on understanding the word, or Adam rewarding Betty) or the similar use of the word in other contexts, although the latter requires more inference and is more prone to error. When Betty has selected a concept to associate with the word (or more precisely, has associated each concept that she knows with a probability that that concept is expressed by this word), she can take the role of speaker. If she wants to communicate "round, red, small" ball, she will choose the word that is most closely associated with that concept, that is, the word that she has now learned from Adam.

Techniques such as the one we have sketched will allow agents to develop a shared lexicon to describe a scene that they can both see. But this is only the beginning. To allow effective communication to proceed, the agents need a shared grammar as well as syntax. They also need to be able to use language not only as a method of transmitting descriptions but also as a way of carrying out actions such as commanding, asking, persuading, and so on. We are only just beginning to discover how to build systems in which agents can evolve such capabilities (Cangelosi & Parisi, 2002), but it would be surprising if great progress were not made in the next decade.

RESOURCES

Societies and Associations

There are three regional societies that promote social simulation and agent-based modeling, each with an annual conference. Every other year, they organize a World Congress together.

- *North American Association for Computational Social and Organization Sciences (NAACSOS).* Web site: http://www.casos.cs.cmu.edu/naacsos/
- *Pacific Asian Association for Agent-Based Approach in Social Systems Sciences (PAAA).* Web site: http://www.paaa.econ.kyoto-u.ac.jp/
- *European Social Simulation Association (ESSA).* Web site: http://essa.eu.org/

You can join these associations for a modest annual membership fee, and they provide a very useful entry to the agent-based modeling research community.

Journals

Research using agent-based modeling appears both in discipline-specific journals and in interdisciplinary journals focusing on social simulation. The two most prominent interdisciplinary journals are the following

- *Journal of Artificial Societies and Social Simulation (JASSS).* This is an online electronic journal, available only on the Web. It is free with no subscription. On the front page, at http://jasss.soc.surrey.ac.uk/JASSS.html, you can sign up to receive an e-mail when each issue is published (four issues per year).
- *Computational and Mathematical Organization Theory (CMOT).* This is a hard-copy journal, with an online, charged-for version at http://springerlink.metapress.com/content/1572-9346/.

Other journals with more than occasional agent-based modeling papers include

- *Artificial Life*
- *Complexity*
- *Computational Economics*
- *Ecology and Society*
- *Emergence: Complexity and Organization*
- *Environment and Planning B*
- *Environmental Modeling and Software*
- *Journal of Economic Dynamics and Control*
- *Physica A: Statistical and Theoretical Physics*
- *Simulation and Gaming*
- *Simulation Modeling Practice and Theory*
- *Social Networks*
- *Social Science Computer Review*

Mailing List and Web Sites

The SIMSOC e-mail distribution list sends out notices of forthcoming conferences, workshops, and jobs of interest to agent-based modelers. You can subscribe to the list or view the list archives at http://www.jiscmail.ac.uk/lists/simsoc.html.

There is an excellent Web site maintained by Leigh Tesfatsion called Agent-Based Computational Economics, which has many well-categorized links to a wide range of literature, at http://www.econ.iastate.edu/tesfatsi/ace.htm.

This book's Web site is at http://cress.soc.surrey.ac.uk/qasss/. There you will find the code for the collectivities model and links to agent-based modeling resources.

GLOSSARY

Agent A computer program, or part of a program, that can be considered to act autonomously and that represents an individual, organization, nation-state, or other social actor.

Analogical model A model that is based on an analogy between the *target* being modeled and the form of the model.

Artificial neural network A computational structure consisting of interconnected units that can be trained to model complex relationships between the inputs applied to the network and its outputs.

Attribute A characteristic or feature of an *agent*. It may be set at the beginning of a simulation, and may alter in value during the run to indicate changes in the agent.

Back propagation (Short for "back propagation of errors.") An algorithm that is used to train *artificial neural networks,* so that when presented with training inputs, the levels of their outputs more closely resemble the desired ones. The error between the actual and desired levels for each output unit are spread back through the network, adjusting the *connection weights* according to the errors.

Boundedly rational There are many situations where making a perfectly rational decision would involve infinite computation or require infinite amounts of information. It is therefore assumed that people are boundedly rational, that is, are limited in the amount of cognitive processing in which they can engage when making decisions.

Buffer An area of computer memory used to store values temporarily, often on a first-in, first-out basis.

Chromosome In a *genetic algorithm,* a chromosome is a set of parameters that defines a proposed solution. The chromosome is often composed of a sequence of binary digits, or of floating point or integer values.

Class In object-oriented programming, a class is a specification of a type of object, showing what attributes and methods instances of that class would have.

Classifier system (Short for "learning classifier system.") First described by Holland (1975), a classifier system consists of a collection of binary rules. A *genetic algorithm* modifies and selects the best rules. The *fitness* of a rule is decided by a *reinforcement learning* technique.

Companion modeling A form of modeling in which models are developed in close association with the people who might be represented in the model and who might benefit from the knowledge and understanding that the model yields.

Connection weight In an artificial neural network, this is the numerical weighting factor applied to an input to a unit. The *back propagation* algorithm modifies the weights as the network is trained.

Control group   In the social and medical sciences, an experiment is typically carried out on two similar groups, one of which receives the treatment while the other, the control group, does not. The outcomes in the two groups are then compared.

Crossover   In evolutionary computation, a method of creating a new *chromosome* from corresponding parts of its parents' chromosomes. Crossover is often used in *genetic algorithms.*

Environment   The simulated surroundings in which an agent is located, possibly including simulated physical elements and other agents.

Equation-based model   A model consisting of one or more equations that relate variables describing a system (Parunak, Savit, & Riolo, 1998).

Fitness   In evolutionary computation, a measure of the adequacy of an individual within an environment. The fitness is used to determine the likelihood that the individual will reproduce and pass parts of its chromosome to the next generation.

Framework   A program or library intended to make it easier to develop agent-based models. The framework provides some standardized components and possibly a basic design for the model.

Genetic algorithm (GA)   A method of simulating evolution. A population of individuals each with a fixed-length *chromosome* is evolved by employing *crossover* and mutation operators and a *fitness* function that determines how likely individuals are to reproduce. GAs perform a type of search in a fitness landscape, aiming to find an individual that has optimum fitness.

Geographical information system (GIS)   A type of database in which a common spatial coordinate system is the primary means of reference. GISs provide facilities for data input, storage, retrieval, and representation; data management, transformation, and analysis; and data reporting and visualization.

Global variable   A variable whose value may be accessed and set throughout a program, rather than only within some restricted context.

Ideal type   An ideal-type model is one that is formed by the one-sided accentuation of one or more characteristics to create a unified analytical construct that abstracts from the variety actually found in concrete social phenomena.

Instantiate   The process of forming an object (in an object-oriented computer language) by following the specifications represented by a *class.* The object is an instance of the class.

Language game   An interaction between two agents consisting of the interchange of simple messages, with the aim of helping to develop a shared understanding of the meaning of the messages.

Machine learning   Algorithms that allow programs to improve their ability to undertake some task by learning from their previous experience.

Message   A symbolic communication between two agents, often represented as a string of characters.

Method   In object-oriented programming, a piece of program code associated with a *class* that performs some function, often in response to a message received by an instance of the class.

Micro-foundation   Assumptions and theories about individual behavior that inform the design of agents. The actions of the agents are expected to lead to the emergence of features that correspond to real-world macro phenomena.

Model   A simplified representation of some social phenomenon. Executing or "running" the model yields a simulation whose behavior is intended to mirror some social process or processes.

Modeling environment   A computer program that allows the user to create, execute, and visualize the outputs of a simulation.

Patch   In NetLogo, a cell of the grid that constitutes the *environment* for a model's agents (called "turtles" in NetLogo).

Phase change   An abrupt change in the state of a system as a whole, consequent on a small change in one variable. By analogy to phase transitions in materials, for example, the change from a solid to a liquid when the material's temperature is raised through its melting point.

Policy   In *reinforcement learning,* a policy maps states of the world to the actions the agent ought to take in those states.

Power law   A relationship between two variables such that one is proportional to a power of the other. If one takes logarithms of each variable, the relationship between the logged variables is linear and can be represented as a straight line on a plot of the two logged variables. Many relationships between variables describing complex systems follow a power law.

Production (rule) system   A problem-solving system consisting of a knowledge base of rules and general facts, a working memory of facts concerning the current case, and an inference engine for manipulating both. The rules are generally of the form, "if [condition], then [action]."

Reinforcement learning   A type of *machine learning* concerned with how an *agent* ought to take actions in an *environment* so as to maximize some long-term reward.

Research question   A question whose answer can be found by carrying out research. It needs to be sufficiently specific that the research has a reasonable likelihood of obtaining a result, but not so specific that the results will be of limited use or generality.

Retrodict   To make predictions about the past. Normal predictions estimate what will happen in the future from the basis of some theory and assumptions; retrodictions use theory and assumptions to estimate a past state. Because the past state can be known empirically through measurements, retrodiction can be used as a method of assessing the validity of the theory and assumptions.

Scale model A model in which the simplifications of reality come mainly from making the model smaller than the target being modeled.

Sensitivity analysis A systematic analysis of changes in simulation results as the model's parameters are changed. Sensitivity analysis is used to assess the extent to which the outcomes are dependent on the precise parameter values that have been assumed.

Simulate To run a model and observe its behavior through time.

Spatially explicit A spatially explicit model is one in which geography is represented within the model, for example, by locating all simulated objects on a grid or other spatial representation. Spatially explicit simulations often use a *geographical information system* to manage the location of objects.

Stylized fact A simplified presentation of an empirical finding that, although broadly true, may ignore particular exceptions. Usually, stylized facts describe whole societies or economies rather than the characteristics of individuals.

Target The social phenomenon or process that is represented by a model.

Toroid A doughnut-shaped object that can be constructed by rotating a circle around an axis external to the circle. Topologically, a toroid is formed by connecting the left and right edges, and then the top and bottom edges, of a rectangle.

Treatment In an *experiment,* the application of some process selectively to the treatment group, while leaving the *control group* unaffected. Changes to the treatment group that are not found in the control group are considered to be due to the treatment.

Validation The process of checking that a model is a good representation of the *target.*

Verification The process of checking that a model conforms to its specification, that is, that it does not include errors, or "bugs."

REFERENCES

Acerbi, A., & Parisi, D. (2006). Cultural transmission between and within generations. *Journal of Artificial Societies and Social Simulation, 9*(1). Retrieved from http://jasss.soc.surrey.ac .uk/9/1/9.html

Ahrweiler, P., & Gilbert, N. (Eds.). (1998). *Computer simulations in science and technology studies.* Berlin: Springer.

Ahrweiler, P., Pyka, A., & Gilbert, N. (2004). Simulating Knowledge Dynamics in Innovation Networks (SKIN). In R. Leombruni & M. Richiardi (Eds.), *Industry and labor dynamics: The agent-based computational economics approach.* Singapore: World Scientific Press.

Albino, V., Carbonara, N., & Giannoccaro, I. (2003). Coordination mechanisms based on cooperation and competition within industrial districts: An agent-based computational approach. *Journal of Artificial Societies and Social Simulation, 6*(4). Retrieved from http:// jasss.soc.surrey.ac.uk/6/4/3.html

Allen, P. M. (1997). *Cities and regions as self-organizing systems.* Amsterdam: Gordon and Breach.

Axelrod, R. (1997a). Advancing the art of simulation in the social sciences. In R. Conte, R. Hegselmann, & P. Terna (Eds.), *Simulating social phenomena* (pp. 21–40). Berlin: Springer.

Axelrod, R. M. (1997b). *The complexity of cooperation: Agent-based models of competition and collaboration.* Princeton, NJ: Princeton University Press.

Axelrod, R. (1997c). The dissemination of culture. *Journal of Conflict Resolution, 41*(2), 203–226.

Axelrod, R. M., & Dawkins, R. (1990). *The evolution of cooperation.* Harmondsworth, UK: Penguin.

Bagnall, A. J., & Smith, G. D. (2005). A multiagent model of the UK market in electricity generation. *Evolutionary Computation, 9*(5), 522–536.

Banzhaf, W. (1998). *Genetic programming: An introduction: On the automatic evolution of computer programs and its applications.* San Francisco: Morgan Kaufmann.

Barabási, A.-L. (2003). *Linked: How everything is connected to everything else and what it means for business, science, and everyday life.* New York: Plume.

Barabási, A.-L., & Albert, R. (1999). Emergence of scaling in random networks. *Science, 286,* 509–512.

Barreteau, O. (2003). Our companion modelling approach. *Journal of Artificial Societies and Social Simulation, 6*(2). Retrieved from http://jasss.soc.surrey.ac.uk/6/2/1.html

Barreteau, O., Bousquet, F., & Attonaty, J.-M. (2001). Role-playing games for opening the black box of multi-agent systems: Method and lessons of its application to Senegal River Valley irrigated systems. *Journal of Artificial Societies and Social Simulation, 4*(2). Retrieved from http://jasss.soc.surrey.ac.uk/4/2/5.html

Barreteau, O., Le Page, C., & D'Aquino, P. (2003). Role-playing games, models and negotiation processes. *Journal of Artificial Societies and Social Simulation, 6*(2). Retrieved from http://jasss.soc.surrey.ac.uk/6/2/10.html

Batten, D., & Grozev, G. (2006). NEMSIM: Finding ways to reduce greenhouse gas emissions using multi-agent electricity modelling. In P. Perez & D. Batten (Eds.), *Complex science for a complex world* (pp. 227–252). Canberra: Australian National University.

Beltratti, A., Margarita, S., & Terna, P. (1996). *Neural networks for economic and financial modelling.* London: International Thomson Computer Press.

Benenson, I., Omer, I., & Hatna, E. (2002). Entity-based modelling of urban residential dynamics: The case of Yaffo, Tel-Aviv. *Environment and Planning B, 29,* 491–512.

Boden, M. A. (1988). *Computer models of mind: Computational approaches in theoretical psychology.* Cambridge, UK: Cambridge University Press.

Boero, R., Castellani, M., & Squazzoni, F. (2004). Micro behavioural attitudes and macro technological adaptation in industrial districts: An agent-based prototype. *Journal of Artificial Societies and Social Simulation, 7*(2). Retrieved from http://jasss.soc.surrey.ac.uk/7/2/1.html

Boero, R., & Squazzoni, F. (2005). Does empirical embeddedness matter? Methodological issues on agent based models for analytical social science. *Journal of Artificial Societies and Social Simulation, 8*(4). Retrieved from http://jasss.soc.surrey.ac.uk/8/4/6.html

Borrelli, F., Ponsiglione, C., Iandoli, L., & Zollo, G. (2005). Inter-organizational learning and collective memory in small firms clusters: An agent-based approach. *Journal of Artificial Societies and Social Simulation, 8*(3). Retrieved from http://jasss.soc.surrey.ac.uk/8/3/4.html

Bourdieu, P. (1986). *Distinction: A social critique of the judgement of taste* (R. Nice, Trans.). New York: Routledge.

Brenner, T. (2001). Simulating the evolution of localised industrial clusters: An identification of the basic mechanisms. *Journal of Artificial Societies and Social Simulation, 4*(3). Retrieved from http://jasss.soc.surrey.ac.uk/4/3/4.html

Brown, L., & Harding, A. (2002). Social modelling and public policy: Application of microsimulation modelling in Australia. *Journal of Artificial Societies and Social Simulation, 5*(4). Retrieved from http://jasss.soc.surrey.ac.uk/5/4/6.html

Bruch, E. E., & Mare, R. D. (2006). Neighborhood choice and neighborhood change. *American Journal of Sociology, 112*(4), 667–709.

Bull, L. (2004). Learning classifier systems: A brief introduction. In L. Bull (Ed.), *Applications of learning classifier systems* (pp. 3–14). New York: Springer.

Bunn, D. W., & Oliveira, F. S. (2001). Agent-based simulation: An application to the new electricity trading arrangements of England and Wales. *Evolutionary Computation, 5*(5), 493–503.

Cangelosi, A., & Parisi, D. (2002). *Simulating the evolution of language.* London: Springer.

Carley, K. M. (1999). On generating hypotheses using computer simulations. *Systems Engineering, 2*(2), 69–77.

Castle, C. J. E., & Crooks, A. T. (2006). *Principles and concepts of agent-based modelling for developing geospatial simulations.* University College London. Retrieved from http://www.casa.ucl.ac.uk/publications/workingPaperDetail.asp?ID=110

Cecconi, F., & Parisi, D. (1998). Individual versus social survival strategies. *Journal of Artificial Societies and Social Simulation, 1*(2). Retrieved from http://jasss.soc.surrey.ac.uk/1/2/1.html

Cederman, L.-E. (1997). *Emergent actors in world politics: How states and nations develop and dissolve.* Princeton, NJ: Princeton University Press.

Chan, N., LeBaron, B., Lo, A., & Poggio, T. (1999). Agent-based models of financial markets: A comparison with experimental markets. *MIT Artificial Markets Project, Paper No. 124, September.* Retrieved from http://ebusiness.mit.edu/research/papers/124%20Poggio%20Lo,%20Agent-based%20Models.pdf

Chang, K.-T. (2004). *Introduction to geographic information systems* (2nd ed.). Boston: McGraw-Hill.

Chatfield, C. (2004). *The analysis of time series: An introduction* (6th ed.). Boca Raton, FL: Chapman & Hall/CRC.

Chattoe, E. (1998). Just how (un)realistic are evolutionary algorithms as representations of social processes? *Journal of Artificial Societies and Social Simulation, 1*(3). Retrieved from http://jasss.soc.surrey.ac.uk/1/3/2.html

Chattoe, E., Saam, N. J., & Möhring, M. (2000). Sensitivity analysis in the social sciences: Problems and prospects. In K. Troitzsch, N. Gilbert, & R. Suleiman (Eds.), *Tools and techniques for social science simulation*. New York: Physica-Verlag.

Chen, Y., & Tang, F. (1998). Learning and incentive-compatible mechanisms for public goods provision: An experimental study. *Jounal of Political Economy, 106,* 633–662.

Clark, W. A. (1991). Residential preferences and neighborhood racial segregation: A test of the Schelling segregation model. *Demography, 28*(1), 1–19.

Cobb, C. W., & Douglas, P. H. (1928). A theory of production. *American Economic Review, 18*(Suppl.), S139–S165.

Cohen, M. D., March, J. G., & Olsen, J. (1972). A garbage can model of organizational choice. *Administrative Science Quarterly, 17*(1), 1–25.

Conte, R., & Castelfranchi, C. (1995). *Cognitive and social action*. London: UCL Press.

D'Aquino, P., Le Page, C., Bousquet, F., & Bah, A. (2003). Using self-designed role-playing games and a multi-agent system to empower a local decision-making process for land use management: The SelfCormas experiment in Senegal. *Journal of Artificial Societies and Social Simulation, 6*(3). Retrieved from http://jasss.soc.surrey.ac.uk/6/3/5.html

Dean, J. S., Gumerman, G. J., Epstein, J. M., Axtell, R. L., Swedland, A. C., Parker, M. T., & McCarroll, S. (1999). Understanding Anasazi culture change through agent-based modeling. In T. Kohler & G. Gumerman (Eds.), *Dynamics in human and primate societies*. Oxford, UK: Oxford University Press.

Deffuant, G. (2006). Comparing extremism propagation patterns in continuous opinion models. *Journal of Artificial Societies and Social Simulation, 9*(3). Retrieved from http://jasss.soc.surrey.ac.uk/9/3/8.html

Deffuant, G., Amblard, F., & Weisbuch, G. (2002). How can extremism prevail? A study based on the relative agreement interaction model. *Journal of Artificial Societies and Social Simulation, 5*(4). Retrieved from http://jasss.soc.surrey.ac.uk/5/4/1.html

Degushi, H. (2004). *Economics as an agent-based complex system*. Tokyo: Springer-Verlag.

Dibble, C., & Feldman, P. G. (2004). The GeoGraph 3D Computational Laboratory: Network and terrain landscapes for RePast. *Journal of Artificial Societies and Social Simulation, 7*(1). Retrieved from http://jasss.soc.surrey.ac.uk/7/1/7.html

Dray, A., Perez, P., Jones, N., Le Page, C., D'Aquino, P., White, I., & Auatabu, T. (2006). The AtollGame experience: From knowledge engineering to a computer-assisted role playing game. *Journal of Artificial Societies and Social Simulation, 9*(1). Retrieved from http://jasss.soc.surrey.ac.uk/9/1/6.html

Dunham, J. B. (2005). An agent-based spatially explicit epidemiological model in MASON. *Journal of Artificial Societies and Social Simulation, 9*(1). Retrieved from http://jasss.soc.surrey.ac.uk/9/1/3.html

Eckel, B. (2005). *Thinking in Java*. Englewood Cliffs, NJ: Prentice Hall.

Edmonds, B. (2006). The emergence of symbiotic groups resulting from skill-differentiation and tags. *Journal of Artificial Societies and Social Simulation, 9*(1). Retrieved from http://jasss.soc.surrey.ac.uk/9/1/10.html

Edmonds, B., & Hales, D. (2003). Replication, replication and replication: Some hard lessons from model alignment. *Journal of Artificial Societies and Social Simulation, 6*(4). Retrieved from http://jasss.soc.surrey.ac.uk/6/4/11.html

Eiben, A. E., & Smith, J. E. (2003). *Introduction to evolutionary computing*. New York: Springer.

Elias, N. (1969). *The civilising process*. Oxford, UK: Blackwell. (Original work published 1939)

Engelbrecht, A. P. (2002). *Computational intelligence: An introduction*. Chichester, UK: Wiley.

Epstein, J. M. (1999). Agent-based computational models and generative social science. *Complexity, 4*(5), 41–60.

Epstein, J. M. (2007). *Generative social science: Studies in agent-based computational modeling.* Princeton, NJ: Princeton University Press.

Epstein, J. M., & Axtell, R. (1996). *Growing artificial societies: Social science from the bottom up.* Washington, DC: Brookings Institution Press.

Erev, I., & Roth, A. E. (1998). Predicting how people play games: Reinforcement learning in experimental games with unique, mixed strategy equilibria. *American Economic Review, 88*(4), 848–881.

Etienne, M. (2003). SYLVOPAST: A multiple target role-playing game to assess negotiation processes in sylvopastoral management planning. *Journal of Artificial Societies and Social Simulation, 6*(2). Retrieved from http://jasss.soc.surrey.ac.uk/6/2/5.html

Etienne, M., Le Page, C., & Cohen, M. (2003). A step-by-step approach to building land management scenarios based on multiple viewpoints on multi-agent system simulations. *Journal of Artificial Societies and Social Simulation, 6*(2). Retrieved from http://jasss.soc.surrey.ac.uk/6/2/2.html

Fagiolo, G., Windrum, P., & Moneta, A. (2006). *Empirical validation of agent-based models: A critical survey* (No. 2006/14). Pisa, Italy: Sant'Anna School of Advanced Studies, Laboratory of Economics and Management.

Farmer, J. D., Patelli, P., & Zovko, L. I. (2005). The predictive power of zero intelligence in financial markets. *Proceedings of the National Academy of Sciences, 102*(6), 2254–2259. Retrieved from http://www.santafe.edu/ ∼ jdf/papers/zero.pdf

Fielding, J., & Gilbert, N. (2000). *Understanding social statistics.* London: Sage.

Fink, E. C., Gates, S., & Humes, B. D. (1998). *Game theory topics: Incomplete information, repeated games and n-player games* (Sage Series on Quantitative Applications in the Social Sciences, Vol. 122). Thousand Oaks, CA: Sage.

Fioretti, G. (2001). Information structure and behaviour of a textile industrial district. *Journal of Artificial Societies and Social Simulation, 4*(4). Retrieved from http://jasss.soc.surrey.ac.uk/4/4/1.html

Flache, A., & Macy, M. (2002). Stochastic collusion and the power law of learning. *Journal of Conflict Research, 46*(5), 629–653.

Forrester, J. W. (1971). *World dynamics.* Cambridge: MIT Press.

Fotheringham, A. S., & O'Kelly, M. E. (1989). *Spatial interaction models: Formulations and applications.* Dordrecht: Kluwer.

Friedman-Hill, E. (2003). *Jess in action: Rule-based systems in Java.* Greenwich, CT: Manning.

Galan, J. M., & Izquierdo, L. (2005). Appearances can be deceiving: Lessons learned re-implementing Axelrod's "evolutionary approach to norms." *Journal of Artificial Societies and Social Simulation, 8*(3). Retrieved from http://jasss.soc.surrey.ac.uk/8/3/2.html

Garson, G. D. (1998). *Neural networks: An introductory guide for social scientists.* London: Sage.

Gaylord, R. J., & D'Andria, L. (1998). *Simulating society: A Mathematica toolkit for modeling socioeconomic behavior.* New York: TELOS/Springer Verlag.

Gilbert, N. (1997). A simulation of the structure of academic science. *Sociological Research Online, 2*(2). Retrieved from http://www.socresonline.org.uk/socresonline/2/2/3.html

Gilbert, N. (2002). Varieties of emergence. In D. Sallach (Ed.), *Agent 2002: Social agents: Ecology, exchange, and evolution* (pp. 41–56). Chicago: University of Chicago and Argonne National Laboratory.

Gilbert, N. (2006). *A generic model of collectivities*. ABModSim 2006, International Symposium on Agent Based Modeling and Simulation. University of Vienna: European Meeting on Cybernetic Science and Systems Research.

Gilbert, N., & Abbott, A. (Eds.). (2005). Social science computation [Special issue]. *American Journal of Sociology, 110*(4). Chicago: University of Chicago Press.

Gilbert, N., & Bankes, S. (2002). Platforms and methods for agent-based modeling. *Proceedings of the National Academy of Sciences, 99*(Suppl. 3), 7197–7198.

Gilbert, N., den Besten, M., Bontovics, A., Craenen, B. G. W., Divina, F., Eiben, A. E., et al. (2006). Emerging artificial societies through learning. *Journal of Artificial Societies and Social Simulation, 9*(2). Retrieved from http://jasss.soc.surrey.ac.uk/9/2/9.html

Gilbert, N., Pyka, A., & Ahrweiler, P. (2001). Innovation networks: A simulation approach. *Journal of Artificial Societies and Social Simulation, 4*(3). Retrieved from http://www.soc.surrey.ac.uk/JASSS/4/3/8.html

Gilbert, N., & Terna, P. (2000). How to build and use agent-based models in social science. *Mind and Society, 1*(1), 57–72.

Gilbert, N., & Troitzsch, K. G. (2005). *Simulation for the social scientist* (2nd ed.). Milton Keynes, UK: Open University Press.

Gimblett, H. R. (2002). *Integrating geographic information systems and agent-based modeling techniques for simulating social and ecological processes*. London: Oxford University Press.

Goolsby, R. (2006). Combating terrorist networks: An evolutionary approach. *Computational and Mathematical Organization Theory, 12*, 7–20.

Grimm, V., & Railsback, S. F. (2005). *Individual-based modeling and ecology*. Princeton, NJ: Princeton University Press.

Gupta, A., & Kapur, V. (2000). *Microsimulation in government policy and forecasting*. Amsterdam: Elsevier.

Hales, D. (2000). Cooperation without space or memory: Tags, groups and the prisoner's dilemma. In S. Moss & P. Davidsson (Eds.), *Multi-agent-based simulation* (pp. 157–166). Berlin: Springer-Verlag.

Hales, D. (2002). Evolving specialisation, altruism and group-level optimisation using tags. In J. S. Sichman, F. Bousquet, & P. Davidsson (Eds.), *Multi-agent-based simulation II* (pp. 26–35). Berlin: Springer-Verlag.

Harding, A. (1996). *Microsimulation and public policy*. New York: Elsevier.

Hegselmann, R., & Krause, U. (2002). Opinion dynamics and bounded confidence models, analysis and simulation. *Journal of Artificial Societies and Social Simulation, 5*(3). Retrieved from http://jasss.soc.surrey.ac.uk/5/3/2.html

Heywood, D. I., Cornelius, S., & Carver, S. (2002). *An introduction to geographical information systems* (2nd ed.). New York: Prentice Hall.

Hodkinson, P. (2002). *Goth identity, style and subculture*. Oxford, UK: Berg.

Holland, J. H. (1975). *Adaptation in natural and artificial systems*. Ann Arbor: University of Michigan Press.

Homer, J. B., & Hirsch, G. B. (2006). System dynamics modeling for public health: Background and opportunities. *American Journal of Public Health, 96*(3), 452–458.

Huberman, B. A., & Glance, N. (1993). Evolutionary games and computer simulations. *Proceedings of the National Academy of Sciences, 90*, 7716–7718.

Hutchins, E., & Hazlehurst, B. (1995). How to invent a lexicon: The development of shared symbols in interaction. In N. Gilbert & R. Conte (Eds.), *Artificial societies*. London: UCL Press.

Izquierdo, S. S., & Izquierdo, L. R. (2006). *The impact on market efficiency of quality uncertainty without asymmetric information.* Paper presented at the Workshop on Agent-Based Models of Consumer Behaviour and Market Dynamics, Guildford, UK.

Jacobsen, C., Bronson, R., & Operations Research Society of America. (1985). *Simulating violators.* Arlington, VA: Operations Research Society of America.

Janssen, M., & Jager, W. (1999). An integrated approach to simulating behavioural processes: A case study of the lock-in of consumption patterns. *Journal of Artificial Societies and Social Simulation, 2*(2). Retrieved from http://jasss.soc.surrey.ac.uk/2/2/2.html

Johnson, P. E. (2002). Agent-based modeling: What I learned from the artificial stock market. *Social Science Computer Review, 20,* 174–186.

Kahneman, D. (2003). Maps of bounded rationality: Psychology for behavioral economics. *American Economic Review, 93*(5), 1449–1475.

Kaldor, N. (1961). Capital accumulation and economic growth. In F. A. Lutz & D. C. Hague (Eds.), *The theory of capital* (pp. 177–222). London: Macmillan.

Kitts, J. A., Macy, M. W., & Flache, A. (1999). Structural learning: Attraction and conformity in task-oriented groups. *Computational & Mathematical Organization Theory, 5*(2), 129–145.

Klemm, K., Eguíluz, V. M., Toral, R., & San Miguel, M. (2003). Global culture: A noise-induced transition in finite systems. *Physcial Review E, 67*(4). Retrived from http://arxiv.org/abs/cond-mat/0205188

Koesrindartoto, D., Sun, J., & Tesfatsion, L. S. (2005). *An agent-based computational laboratory for testing the economic reliability of wholesale power market designs.* Paper presented at the Energy, Environment, and Economics in a New Era conference, San Francisco.

Koza, J. R. (1992). *Genetic programming.* Cambridge: MIT Press.

Koza, J. R. (1994). *Genetic programming 2.* Cambridge: MIT Press.

Laird, J. E., Newell, A., & Rosenbloom, P. S. (1987). Soar: An architecture for general intelligence. *Artificial Intelligence, 33*(1), 1–64.

LeBaron, B., Arthur, W. B., & Palmer, R. (1999). Time series properties of an artificial stock market. *Journal of Economic Dynamics and Control, 23,* 1487–1516.

Liebrand, B. G., Nowak, A., & Hegselmann, R. (Eds.). (1998). *Computer modeling of social processes.* London: Sage.

Link, J., & Fröhlich, P. (2003). *Unit testing in Java: How tests drive the code.* San Francisco: Morgan Kaufmann.

Lorenz, J. (2006). Consensus strikes back in the Hegselmann-Krause model of continuous opinion dynamics under bounded confidence. *Journal of Artificial Societies and Social Simulation, 9*(1). Retrieved from http://jasss.soc.surrey.ac.uk/9/1/8.html

Luke, S., Cioffi-Revilla, C., Panait, L., Sullivan, K., & Balan, G. (2005). MASON: A Java multi-agent simulation environment. *Simulation: Transactions of the Society for Modeling and Simulation International, 81*(7), 517–527.

Macy, M., & Flache, A. (2002). Learning dynamics in social dilemmas. *Proceedings of the National Academy of Sciences, 99*(Suppl. 3), 7229–7236.

Macy, M., & Willer, R. (2002). From factors to actors: Computational sociology and agent-based modeling. *Annual Review of Sociology, 28,* 143–166.

Malerba, F., Nelson, R., Orsenigo, L., & Winter, S. G. (2001). History-friendly models: An overview of the case of the computer industry. *Journal of Artificial Societies and Social Simulation, 4*(3). Retrieved from http://jasss.soc.surrey.ac.uk/4/3/6.html

Marengo, L. (1992). Coordination and organizational learning in the firm. *Journal of Evolutionary Economics, 2*(3), 313–326.

McKeown, G., & Sheehy, N. (2006). Mass media and polarisation processes in the bounded confidence model of opinion dynamics. *Journal of Artificial Societies and Social Simulation, 9*(1). Retrieved from http://jasss.soc.surrey.ac.uk/9/1/11.html

Merton, R. K. (1968). *Social theory and social structure* (Enl. ed.). New York: Free Press.

Meyer, M., & Hufschlag, K. (2006). A generic approach to an object-oriented learning classifier system library. *Journal of Artificial Societies and Social Simulation, 9*(3). Retrieved from http://jasss.soc.surrey.ac.uk/9/3/9.html

Miles, R. (2006). *Learning UML 2.0.* Sebastopol, CA: O'Reilly.

Mitton, L., Sutherland, H., & Weeks, M. J. (2000). *Microsimulation modelling for policy analysis: Challenges and innovations.* Cambridge, UK: Cambridge University Press.

Mookherjee, D., & Sopher, B. (1994). Learning behavior in an experimental matching pennies game. *Games and Economic Behavior, 7,* 62–91.

Mookherjee, D., & Sopher, B. (1997). Learning and decision costs in experimental constant sum games. *Games and Economic Behavior, 19,* 97–132.

Moss, S. (2002). Policy analysis from first principles. *Proceedings of the National Academy of Sciences, 99*(Suppl. 3), 7267–7274.

Mulkay, M. J., & Turner, B. S. (1971). Over-production of personnel and innovation in three social settings. *Sociology, 5*(1), 47–61.

Niemeyer, P., & Knudsen, J. (2005). *Learning Java* (3rd ed.). Sebastopol, CA: O'Reilly.

Nilsson, N. (1998). *Artificial intelligence: A new synthesis.* San Francisco: Morgan Kaufmann.

North, M. J. (2001). Multi-agent social and organizational modeling of electric power and natural gas markets. *Computational and Mathematical Organization Theory, 7,* 331–337.

North, M. J., Collier, N. T., & Vos, J. R. (2006). Experiences creating three implementations of the repast agent modeling toolkit. *ACM Transactions on Modeling and Computer Simulation, 16*(1), 1–25.

Orcutt, G., Merz, J., & Quinke, H. (1986). *Microanalytic simulation models to support social and financial policy.* Amsterdam: North-Holland.

O'Sullivan, D., & Macgill, J. (2005). *Modelling urban residential neighbourhood dynamics.* Paper presented at the Workshop on Modelling Urban Social Dynamics, University of Surrey, Guildford, UK.

Pajares, J., Hernández-Iglesias, C., & López-Paredes, A. (2004). Modelling learning and R&D in innovative environments: A cognitive multi-agent approach. *Journal of Artificial Societies and Social Simulation, 7*(2). Retrieved from http://jasss.soc.surrey.ac.uk/7/2/7.html

Pancs, R., & Vriend, N. J. (2004). *Schelling's spatial proximity model of segregation revisited.* International Workshop on "Agent-Based Models for Economic Policy Design" (ACEPOL05). Retrieved January 20, 2007, from http://www.wiwi.uni-bielefeld.de/~ dawid/acepol/Downloads/Paper/Vriend.pdf

Papert, S. (1980). *Mindstorms: Children, computers, and powerful ideas.* New York: Basic Books.

Parunak, H. V. D., Savit, R., & Riolo, R. L. (1998, July). Agent-based modeling vs. equation-based modeling: A case study and users' guide. In J. S. Sichman, R. Conte, & N. Gilbert (Eds.), *Proceedings of Multi-Agent Systems and Agent-Based Simulation* (pp. 10–25). Paris: Springer.

Perfors, A. (2002). Simulated evolution of language: A review of the field. *Journal of Artificial Societies and Social Simulation, 5*(2). Retrieved from http://jasss.soc.surrey.ac.uk/5/2/4.html

Phillips, A. W. (1950). Mechanical models in economic dynamics. *Economica, 17*(67), 283–305.

Powell, W. W., White, D. R., Koput, K. W., & Owen-Smith, J. (2005). Network dynamics and field evolution: The growth of inter-organizational collaboration in the life sciences. *American Journal of Sociology, 110*(4), 1132–1205.

Pujol, J. M., Flache, A., Delgado, J., & Sangüesa, R. (2005). How can social networks ever become complex? Modelling the emergence of complex networks from local social exchanges. *Journal of Artificial Societies and Social Simulation, 8*(4). Retrieved from http://jasss.soc.surrey.ac.uk/8/4/12.html

Railsback, S. F., Lytinen, S. L., & Jackson, S. K. (2006). Agent-based simulation platforms: Review and development recommendations. *Simulation: Transactions of the Society for Modeling and Simulation International, 82*(9), 609–623. Available: http://www.humboldt .edu/ ~ ecomodel/documents/ABMPlatformReview.pdf

Ramanath, A. M., & Gilbert, N. (2004). Techniques for the construction and evaluation of participatory simulations. *Journal of Artificial Societies and Social Simulation, 7*(4). Retrieved from http://jasss.soc.surrey.ac.uk/7/4/1.html

Raney, B., Cetin, N., Vollmy, A., & Nagel, K. (2002). *Large scale multi-agent transportation simulations.* Paper presented at the Proceedings of the Annual Congress of the European Regional Science Association (ERSA), Dortmund, Germany. Available: http://www .inf.ethz.ch/nagel/papers

Redmond, G., Sutherland, H., & Wilson, M. (1998). *The arithmetic of tax and social security reform: A user's guide to microsimulation methods and analysis.* Cambridge, UK: Cambridge University Press.

Reynolds, C. W. (1987). Flocks, herds, and schools: A distributed behavioral model. *Computer Graphics, 21*(4), 25–34.

Richiardi, M., Leombruni, R., Saam, N., & Sonnessa, M. (2006). A common protocol for agent-based social simulation. *Journal of Artificial Societies and Social Simulation, 9*(1). Retrieved from http://jasss.soc.surrey.ac.uk/9/1/15.html

Riolo, R. L., Cohen, M. D., & Axelrod, R. (2001). Evolution of cooperation without reciprocity. *Nature, 411,* 441–443.

Roth, A. E., & Erev, I. (1995). Learning in extensive-form games: Experimental data and simple dynamic models in the intermediate term. *Games and Economic Behavior, 8,* 164–212.

Rouchier, J. (2003). Re-implementation of a multi-agent model aimed at sustaining experimental economic research: The case of simulations with emerging speculation. *Journal of Artificial Societies and Social Simulation, 6*(4). Retrieved from http://jasss.soc.surrey.ac.uk/6/ 4/7.html

Sakoda, J. M. (1971). The checkerboard model of social interaction. *Journal of Mathematical Sociology, 1*(1), 119–131.

Sallans, B., Pfister, A., Karatzoglou, A., & Dorffner, G. (2003). Simulation and validation of an integrated markets model. *Journal of Artificial Societies and Social Simulation, 6*(4). Retrieved from http://jasss.soc.surrey.ac.uk/6/4/2.html

Sawyer, R. K. (2004). Social explanation and computational simulation. *Philosophical Explorations, 7*(3), 219–231.

Schelling, T. C. (1971). Dynamic models of segregation. *Journal of Mathematical Sociology, 1,* 143–186.

Schelling, T. C. (1978). *Micromotives and macrobehavior.* New York: Norton.

Schweitzer, F. (2003). *Brownian agents and active particles.* New York: Springer-Verlag.

Scott, J. (2000). *Social network analysis* (2nd ed.). London: Sage.

Simmel, G. (1907). Fashion. *International Quarterly, 10,* 130–155.

Simon, H. A. (1957). A behavioral model of rational choice. In H. A. Simon (Ed.), *Models of man.* New York: Wiley.

Slanina, F. (2000). Social organization in the minority game model. Retrieved from http://arxiv .org/abs/cond-mat/0006098

Squazzoni, F., & Boero, R. (2002). At the edge of variety and coordination: An agent-based computational model of industrial districts. *Journal of Artificial Societies and Social Simulation, 5*(1). Retrieved from http://jasss.soc.surrey.ac.uk/5/1/1.html

Stauffer, D., Sousa, A., & Schulze, C. (2004). Discretized opinion dynamics of the Deffaunt model on scale-free networks. *Journal of Artificial Societies and Social Simulation, 7*(3). Retrieved from http://jasss.soc.surrey.ac.uk/7/3/7.html

Steels, L. (2004). The evolution of communication systems by adaptive agents. In E. Alonso, D. Kudenko, & D. Kazakov (Eds.), *Adaptive agents and multi-agent systems* (Vol. 2636, pp. 125–140). Berlin: Springer Verlag.

Sterman, J. (2000). *Business dynamics: Systems thinking and modeling for a complex world.* Boston: Irwin McGraw-Hill.

Strader, T. J., Lin, F., & Shaw, M. J. (1998). Simulation of order fulfillment in divergent assembly supply chains. *Journal of Artificial Societies and Social Simulation, 1*(2). Retrieved from http://jasss.soc.surrey.ac.uk/1/2/5.html

Sun, R. (2006). The CLARION cognitive architecture: Extending cognitive modeling to social simulation. In R. Sun (Ed.), *Cognition and multi-agent interaction: From cognitive modeling to social simulation.* New York: Cambridge University Press.

Sun, R., & Naveh, I. (2004). Simulating organizational decision-making using a cognitively realistic agent model. *Journal of Artificial Societies and Social Simulation, 7*(3). Retrieved from http://jasss.soc.surrey.ac.uk/7/3/5.html

Sutton, R. S., & Barto, A. G. (1998). *Reinforcement learning: An introduction.* Cambridge: MIT Press.

Taatgen, N., Lebiere, C., & Anderson, J. (2006). Modeling paradigms in ACT-R. In R. Sun (Ed.), *Cognition and multi-agent interaction: From cognitive modeling to social simulation.* Cambridge, UK: Cambridge University Press.

Tesfatsion, L., & Judd, K. (2006). *Handbook of computational economics* (Vol. 2). Amsterdam: North-Holland.

Thorngate, W. (2000). Teaching social simulation with Matlab. *Journal of Artificial Societies and Social Simulation, 3*(1). Retrieved from http://www.soc.surrey.ac.uk/JASSS/3/1/forum/1.html

Thornton, S. (1995). *Club cultures: Music, media and subcultural capital.* Cambridge, MA: Polity.

Tobias, R., & Hofmann, C. (2004). Evaluation of free Java-libraries for social-scientific agent based simulation. *Journal of Artificial Societies and Social Simulation, 7*(1). Retrieved from http://jasss.soc.surrey.ac.uk/7/1/6.html

Troitzsch, K. G. (2004). *Validating simulation models.* 18th European Simulation Multiconference, SCS Europe.

Vogt, P. (2005). The emergence of compositional structures in perceptually grounded language games. *Artificial Intelligence, 167*(1–2), 206–242.

Vogt, P., & Coumans, H. (2003). Investigating social interaction strategies for bootstrapping lexicon development. *Journal of Artificial Societies and Social Simulation, 6*(1). Retrieved from http://jasss.soc.surrey.ac.uk/6/1/4.html

Waterman, D. A., & Hayes-Roth, F. (1978). *Pattern-directed inference systems.* Orlando, FL: Academic Press.

Watts, D. (1999). Network dynamics and the small world phenomenon. *American Journal of Sociology, 105*(2), 493–527.

Watts, D., & Strogatz, S. (1998). Collective dynamics of "small-world" networks. *Nature, 393,* 440–442.

Widdicombe, S., & Wooffitt, R. C. (1990). "Being" versus "doing" punk (etc): On achieving authenticity as a member. *Journal of Language and Social Psychology, 9,* 257–277.

Wilensky, U. (1998). NetLogo segregation model. Evanston, IL: Northwestern University, Center for Connected Learning and Computer-Based Modeling. Retrieved from http://ccl.northwestern.edu/netlogo/models/Segregation

Wilensky, U. (1999). *NetLogo*. Evanston, IL: Northwestern University, Center for Connected Learning and Computer-Based Modeling. Retrieved from http://ccl.northwestern.edu/netlogo

Wilensky, U. (2005). *NetLogo wolf sheep predation (system dynamics) model*. Evanston, IL: Northwestern University, Center for Connected Learning and Computer-Based Modeling. Retrieved from http://ccl.northwestern.edu/netlogo/models/WolfSheepPredation(System Dynamics)

Wooldridge, M., & Jennings, N. R. (1995). Intelligent agents: Theory and practice. *Knowledge Engineering Review, 10*(2), 115–152.

Wray, R. E., & Jones, R. M. (2006). Considering Soar as an agent architecture. In R. Sun (Ed.), *Cognition and multi-agent interaction: From cognitive modeling to social simulation.* Cambridge, UK: Cambridge University Press.

Ye, M., & Carley, K. M. (1995). Radar-Soar: Towards an artificial organization composed of intelligent agents. *Journal of Mathematical Sociology, 20*(2–3), 219–246.

Zhang, J. (2004). A dynamic model of residential segregation. *Journal of Mathematical Sociology, 28*(3), 147–170.

INDEX

ABOUT THE AUTHOR

Nigel Gilbert is Professor of Sociology at the University of Surrey, Guildford, England. He is the author or editor of 21 books and many academic papers and is the editor of the *Journal of Artificial Societies and Social Simulation.* His current research focuses on the application of agent-based models to understanding social and economic phenomena, especially the emergence of norms, culture, and innovation. He teaches social research methods and computational sociology. He obtained a doctorate in the sociology of scientific knowledge in 1974 from the University of Cambridge and has subsequently taught at the universities of York and Surrey in England. He is one of the pioneers in the field of social simulation and is past president of the European Social Simulation Association. He is an academician of the UK Academy of Social Sciences and a fellow of the Royal Academy of Engineering.